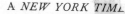

Ja

A *NEW YORK TIME* ~~EDITOR'S CHOICE~~

"A comprehensive _____ ___ ___ __ ___ world's most my-
terious creatures. _____ ___ collected anguilline myths, lore, and
recipes from all over the world."   —*The Economist*

"Enthralling. . . . The eel's story is remarkable, and so are Prosek's tales of
eel people."   —*New Scientist*

"James Prosek sets out to explore the life of an animal that he calls 'timeless
and vital, a metaphor for the resilience of life itself'. He is an ideal guide
to this world. . . . His knowledge and abiding interest permeate the book."
                                                    —*The Seattle Times*

"A riveting synthesis of cultural, geographical, and botanical sleuthing."
                                                    —*Publishers Weekly*

"[Prosek] does well in interweaving the mythological and the personal.
The author is also a diligent natural historian, keen to the greater landscape. He vividly evokes a bleached-white coral path reflecting the moonlight on Pohnpei, and an eel catcher on the Delaware River. . . . A warm,
enrapturing paean to the totemic potency of eels." —*Kirkus Reviews*

"Entertaining. . . . Prosek's writing is fluid and relaxed." —*Washington Post*

"It is a great relief on this deceptively familiar earth to be reminded that the
natural world goes on replete with mysteries. Few are so intertwined with
our terrestrial and marine preconceptions as the great family of eels. James
Prosek explores their astonishing lives in service to our proper awe of nature. This is a delightful work with the urgency of a good detective story."
                        —Thomas McGuane, author of *Driving on the Rim*

"An engagingly written account. . . . Readers interested in anthropology and folklore, fishing, and natural history will also enjoy this volume, which is enhanced by the author's etchings." —*Library Journal*

"I loved it! A beautiful adventure story of one of the most widespread and least-known but ecologically important fish. Prosek combines their amazing biology with adventure and politics. We now have to wonder if these, once some of the most common of fish, are going the way of the cod, the passenger pigeon, and the bison." —Bernd Heinrich, author of *Summer World*

"The tale of Ray Turner, a man who still fishes for eels the traditional way with a hand-built weir, is at the heart of the book, tying the mythology, the mystery, and the commerce of eels together into his story." —*Booklist*

"Author and painter James Prosek has pulled off a neat trick with his latest book: he's managed to transform the lowly, slimy eel (which is a fish, by the way) into an object of fascination for the reader. *Eels* is an excellent read." —*Forbes*

"In a captivating account, Prosek takes readers through the scientific, cultural, and culinary worlds of eels. . . . Prosek gives readers a new appreciation for eels, whose populations are dropping precipitously." —*Science News*

"The fifty-million-year-old species wouldn't seem a likely subject for a riveting natural history book—it is covered in slime, after all—but Prosek pulls it off, thanks mostly to the rabid eel aficionados he digs up." —*Outside* magazine

"No person with even a passing interest in the natural world should be without *Eels*." —*Connecticut Post*

"In *Eels*, naturalist James Prosek travels and interviews leading scientists worldwide to examine the *Anguilla* genus. . . . Prosek considers the cultural and economic value we attach to it, interweaving historic vignettes from Aristotle's interest in the origins of European eels to Sigmund Freud's nineteenth-century search for their testes. Prosek visits New Zealand, where the Maori revere the large endemic longfin eel, *Anguilla dieffenbachii* as a religious symbol, which they believe can bark like a dog and scream like a baby. . . . A solid introduction to global *Anguilla* species."

—*Nature* magazine

"Prosek has made his reputation as a kind of underwater Audubon. His trout watercolors, collected in a book when he was still an undergraduate at Yale, bear those particular, exciting hues that still-living fish possess—a quality that fish-catchers cherish and everyday fish-eaters couldn't care less about. As *Eels* demonstrates, Prosek is every bit as good a writer as a painter. Perhaps this is because both his art and writing draw their inspiration from a similar challenge: to express the ineffable, fading aspect of the natural world in the industrialized era, the feeling of bright colors slipping through your fingers. It is this quality that makes *Eels* much more than a fish book. It is an impassioned defense of nature itself, rescued from the tired rhetoric of 1970s-style environmentalism by good, honest shoe-leather reporting. And yet it contains the untainted germ of Age-of-Aquarius eco-consciousness by centering on an essential question: Does a tidy scientific analysis of a creature really tell us all we need to know, or are there numinous qualitites to every life-form that require a different kind of meditation?"

—Paul Greenberg, *New York Times Book Review*

Eels

An **Exploration**, from New Zealand to the Sargasso,
of the World's Most **Mysterious Fish**

James Prosek

HARPER 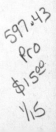 PERENNIAL

NEW YORK • LONDON • TORONTO • SYDNEY • NEW DELHI • AUCKLAND

HARPER ● PERENNIAL

A hardcover edition of this book was published in 2010 by HarperCollins Publishers.

HarperCollins books may be purchased for educational, business, or sales promotional use. For information please write: Special Markets Department, HarperCollins Publishers, 10 East 53rd Street, New York, NY 10022.

FIRST HARPER PERENNIAL EDITION PUBLISHED 2011.

Designed by Leah Carlson-Stanisic

Library of Congress Cataloging-in-Publication Data has been applied for.

ISBN 978-0-06-056612-8

11 12 13 14 15   OV/RRD   10 9 8 7 6 5 4 3 2 1

Thousands passed the lighthouse that night, on the first lap of a far sea journey—all the silver eels, in fact, that the marsh contained. And as they passed through the surf and out to sea, so also they passed from human sight and almost from human knowledge.

—RACHEL CARSON, 1941

Contents

Introduction: *A Transition in Mind to Eels*   *xi*

One   A Mysterious Fish *1*

Two   To the Sargasso *11*

Three   Eels in Maoridom *23*

Four   More Tales of *Taniwha* *91*

Five   The First Taste of Freshwater *115*

Six   Into the Ocean *131*

Seven   Where Eels Go to Die *141*

Eight   Eel Weir Hollow *167*

Nine   The Lasialap of U *189*

Ten   Obstacles in Their Path *251*

Eleven   Still in the Hunt *273*

Acknowledgments   *281*

A TRANSITION

In Mind to Eels

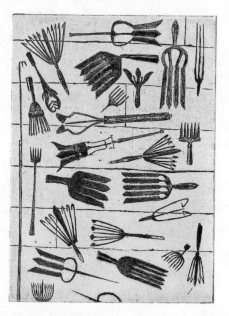

My wall of eel spears

The eel is not an easy fish to like. It doesn't have the beauty of the trout or the colors of the sunfish. As kids, my friends and I caught eels by accident while casting for something else. Unable to grip their muscular, slimy bodies, we stunned them with a smack on the riverbank, then pinned them down with our heels just long enough to remove our hooks from their mouths. We threw them back in the water and watched, astonished, as they swam away.

Eels pulsed through my New England upbringing, there and then not there. Something about them made me curious even if I could not name it. My friend Joe Haines, an old game warden, used to cook them, big eels, trapped unwittingly as they tried to pass through the dam on the Saugatuck Reservoir on their fall migration to the sea. I asked Joe where the eels were going. The answer was always the same—the Sargasso Sea. But where was that? Far away. There was something mysterious in the idea that this fish, which swam in the waters where I waded barefoot in summer, was born thousands of miles away in the blue ocean.

My mother told me that her earliest memory as a girl growing up in Trieste, Italy, was of seeing eels' heads chopped off by the fishmonger in the markets. Why did the eel and the snake, these minimal creatures, make such lasting impressions on our psyches? I stumbled on eels in the photography of Cartier-Bresson, in the paintings of Manet and Leonardo (with orange slices in the Last Supper), in the novels of Günter Grass and Graham Swift. When I started fishing for striped bass we used them for bait, casting them off the beach on chill fall nights where surf spray mingled with stars, on Cuttyhunk Island and Martha's Vineyard.

At the time I had thought these strange travelers were unique to North America and Europe (the European eel also spawns in the Sargasso Sea). I had not known that there were other species of eels around the world that made migrations from the rivers of other continents to spawn in other oceans. A friend who'd lived in New Zealand told me that there were giant freshwater eels in the islands' rivers and that these eels were important to the culture of the indigenous Maori—symbols of the movement of water, synonyms of the phallus, sacred guardians, and monster seducers. He said the Maori kept eels in ponds and fed them by hand, that these fish were known to live for hundreds of years and grow over six feet in length. A fish that once had been an accident on my hook had begun to wedge itself persistently in the crevices of my imagination—a thread that tied the ocean and the rivers together and made me feel like the world was held by one interconnected system of beauty, magic, and mystery.

One day I stood at the edge of a lake in Tuscany with my friend and editor, Larry Ashmead, now retired from Harper-Collins. We looked out across the surface of the water below a Tuscan hill town, and he asked me what the stakes in the mud were marking (the town was Cortona, the lake Trasimeno). I told him that they were traps for eels, fish that were born thousands of miles away in a warm eddy in the ocean called the Sargasso Sea. He said he didn't believe me. When I got home there was a package with a photocopy of an essay by Rachel Carson on eels, and a note from Larry that said he thought the story of this fish might make a good book.

Eels

chapter one

A MYSTERIOUS FISH

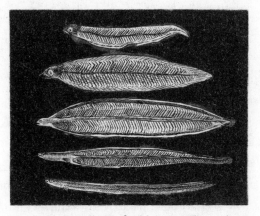

Metamorphosis of eel larva to glass eel

Conjecture about what an eel is exactly, or where its place is in the tree of life, has racked the brains of more than a few naturalists. Its limbless elongated body led some to believe it was related to the snake. The Greek naturalist and poet Oppian wrote in the second century A.D.: "Nothing more is known, than what people repeat about the loves of Roman eels and snakes. Some say that they pair, that, full of eagerness, drunk with desire, the Roman eel comes out of the sea to go and meet her mate." As late as 1833 Jerome V. C. Smith wrote in his

Natural History of the Fishes of Massachusetts: "On the whole, we view the eel in the light of a water-serpent, being the connecting link between purely aquatic and amphibious reptiles." The eel, however, is a fish, though it is a fish like no other.

The freshwater eel, of the genus *Anguilla*, evolved more than fifty million years ago, giving rise to fifteen separate species. Most migratory fish, such as salmon and shad, are anadromous, spawning in freshwater and living their adult lives in salt water. The freshwater eel is one of the few fishes that does the opposite, spawning in the sea and spending its adulthood in lakes, rivers, and estuaries—a life history known as catadromy (in Greek *ana*- means "up" and *cata*- means "down," the prefixes suggesting the direction the fish migrates to reproduce).* But among catadromous fishes, the eel is the only one that travels to the depths of the oceans so far offshore.

"Salmon," Mike Miller, an eel scientist, told me, "can imprint on a river system. They are born in the river system, they go out in the ocean, and they come back to the same river—it's not that bloody

---

* The fish of this book is a member of the family Anguillidae and the order Anguilliformes, which includes the moray and the conger eels (fish that spend their entire lives in the sea). The freshwater eel is not related to other eel-like fish, such as electric eels and lamprey eels, which are not true eels but are evolutionarily quite distinct. Throughout the book I sometimes refer to the catadromous freshwater eel as an anguillid eel. The various species of freshwater eels occupy rivers and streams on the west coast of Africa, Madagascar, India, Indonesia, Australia, China, Korea, Japan, New Zealand, and other islands in Polynesia, Micronesia, and Melanesia (with unique spawning areas in the Indo-Pacific). Curiously, there are no native eels in rivers of the Pacific coast of the Americas. A new, sixteenth species of freshwater eel, *Anguilla luzonensis*, was recently discovered in northern Luzon Island, Philippines.

hard to do. In the case of the eel, you're born in the open ocean. You can't see anything around you except blue water. It's just blue water until they come to the coastal areas, where they first enter estuaries and streams at random. And then, ten to thirty years later when they leave the river, they have to swim all the way out to the same place in the ocean again. And how do they do that?"

The American and European freshwater eels (*Anguilla rostrata* and *Anguilla anguilla*) emerge from eggs suspended in the ocean—specifically, the western part of a subtropical gyre in the North Atlantic somewhere east of Bermuda called the Sargasso Sea.* The only reason scientists know this is that baby eels in their larval stage, called *leptocephali*, have been found drifting near the surface of the ocean thousands of miles from any shore. No one has ever been able to find a spawning adult or witness a freshwater eel spawning in the wild. For eel scientists, solving the mystery of eel reproduction remains a kind of holy grail.

Wherever eels are born, they're relentless in their effort to return to their oceanic womb. I can tell you this from personal experience because I've tried to keep them in a home aquarium. The morning after the first night of my attempt, I found eels slithering around the floor of my kitchen and living room. After

---

* The Sargasso Sea is named after sargassum weed, a prevalent seaweed that floats on its surface. One story has it that the name *sargassum* is from the Portuguese word for "wild grape," *sargaçao*, and describes the weed's bulbous floats. The Sargasso covers two million square miles of the Atlantic and was notorious among sailors for being calm and quiet, like an ocean desert. At the heart of the Bermuda Triangle, the Sargasso became synonymous with lost and abandoned vessels and mystery.

securing a metal screen over the tank with heavy stones, I was able to contain them, but soon they were rubbing themselves raw against the screen. Then one died trying to escape via the filter outflow. When I screened the outflow, they banged their heads against the glass until they had what appeared to be seizures and died. That's when I stopped trying to keep eels.

Eels are wondrous in their ability to move. They're often found in lakes, ponds, and postholes with no visible connection to the sea, leaving the inquisitive shaking their heads. On wet nights, eels are known to cross over land from a pond to a river, or over an obstruction, by the thousands, using each other's moist bodies as a bridge. Young eels can climb moss-covered vertical walls, forming a braid with their bodies. Farmers in Normandy say that eels will leave rivers on spring nights and find their way to vegetable patches to feed on peas.

The yearly journeys millions of adult eels make from rivers to oceans must be among the greatest unseen migrations of any creature on the planet. In the course of these journeys, which span thousands of miles, eels face a long list of dangers: hydroelectric dams, river diversions, pollution, disease, predation (by striped bass, beluga whales, and cormorants, among others), fishing by humans, and changes in ocean currents or temperature structure due to global warming, which may confound eels during their migrations.

From Aristotle through Pliny, Walton, and Linnaeus, great naturalists through time have put forward various theories as to how eels make new eels—that the young emerged from the mud (Aristotle); that eels are bred from a particular dew that falls in the months of May and June (Walton); that they multiplied by rubbing themselves against rocks (Pliny); that they were viviparous, bearing live young rather than laying eggs (Linnaeus). One problem was that no one could identify sperm or eggs in eels. Over a forty-year period in the late 1700s, at the famous eel fishery at Comacchio, Italy, the biologist Spallanzani calculated that more than 152 million adult migratory eels had been caught and cleaned, not one of which was found to be pregnant. No one could say for sure whether eels even had gender, because no one could identify their reproductive organs.*

In the late nineteenth century, a young medical student named Sigmund Freud was assigned by his professor Carl Claus to investigate what had been postulated to be the testes of the male eel. During several months at the Zoological Experimental Station in Trieste, Italy, Freud dissected more than four hundred eels, looking for loops of white matter festooned in the body cavity. His 1877 paper on eels, "Observations on the Form and the Fine Structure of Looped Organs of the Eel, Organs Considered as Testes," was Freud's first published work. One can't help but wonder if the time he spent that summer in his twentieth year dissecting eels played some role in the development of

---

* The sex organs of eels mature only after the adults leave the mouths of rivers for their oceanic spawning grounds.

his later psychosexual theories.* Needless to say, the testes of the eel would not be confirmed until 1897, when a sexually mature male eel was caught in the Straits of Messina between mainland Italy and the island of Sicily.

The larval stage of the eel bears little resemblance to the adult—it is a tiny, transparent creature with a thin head, a body shaped like a willow leaf, and outward-pointing teeth. The larvae of freshwater eels were originally thought to be that of a separate genus of fish, first described as *Leptocephalus breverostris* in 1856 by the German naturalist Kaup after a specimen from the Mediterranean Sea (the common name for eel larvae today, *leptocephali*, remains as a relic of this misnomer). Then in 1896 two Italian biologists, Grassi and Calandruccio, watched a *Leptocephalus breverostris* metamorphose in a tank into an eel—the most convincing evidence to support the theory that freshwater eels were born in salt water. Still, though some speculated that eels spawned in the Mediterranean, no one had yet dreamed that freshwater eels from Europe hatched in the middle of the Atlantic Ocean.

In 1904, Johannes Schmidt, a young Danish fisheries biologist, got a job aboard the *Thor*, a Danish research vessel, studying the breeding habits of food fishes such as cod and herring.

---

* A few of Freud's biographers have actually suggested this (see *The Life and Work of Sigmund Freud*, vol. 1, by Ernest Jones). At the time Freud was pining over a lost love. In a letter to his friend Eduard Silberstein he wrote about the beauty of the Italian woman, and then wryly noted, "It is unfortunately forbidden to dissect humans."

One day in the summer of that year, a tiny fish larva showed up in one of their fine-mesh trawls, west of the Faroe Islands in the Atlantic Ocean. Based on the number of vertebrae, 115, and the hypural bones at the end of the vertebral column, Schmidt identified the larva as that of the European eel, *Anguilla anguilla,* the first to be recognized as such outside of the Mediterranean Sea.[*]

A year before, Schmidt had made an auspicious betrothal to the heiress to the Carlsberg Brewery, a Danish company that had donated generously to marine research. Outfitted with schooners capable of ocean crossings, towing fine-mesh nets and catching hundreds of larvae, he was able to show that the farther from the European coast he went, the smaller the eel larvae became. After almost two decades of cruising the Atlantic, Schmidt was able to assert that both the American and European eels spawned somewhere in the southwestern part of the Atlantic—the Sargasso Sea. "No other instance is known among fishes of a species requiring a quarter of the circumference of the globe to complete its life history," Schmidt wrote in

---

[*] Schmidt later wrote of this early discovery: "I had then only a slight idea of the extraordinary difficulties offered by this problem. . . . The task grew from year to year to an unimaginable degree. In fact, it necessitated cruises of investigation from America to Egypt and from Iceland to the Cape Verde Islands." Eel larvae had been collected in the ocean prior to Schmidt's discovery, but before anyone knew they were the larvae of the eel. He found two small specimens of *leptocephali* larvae in the collections of the Zoological Museum in Copenhagen, captured fifty years before.

1923. "Larval migrations of such extent and duration . . . are altogether unique in the animal kingdom."*

Schmidt and his colleagues went on to search for the spawning areas of other freshwater eel species in the Indo-Pacific region, but with limited success. No other discoveries of anguillid eel-spawning areas came until 1991, when an expedition led by Katsumi Tsukamoto of the Ocean Research Institute in Tokyo found the spawning grounds of the Japanese eel. Japanese scientists had searched unsuccessfully for the spawning area of *Anguilla japonica* for six decades. But on this particular new-moon night in the Pacific Ocean, in the Philippine Sea to the west of Guam, Tsukamoto and the other scientists on board netted the smallest larvae of that species that had ever been collected and finally determined the location of the spawning area of the Japanese freshwater eel. Still, they had not captured any adults.

Mike Miller, then a graduate student from the University of Maine, Orono, who was on the fateful expedition, described what it was like to look for spawning eels in the open ocean. "You could be fifty meters away and not find anything," he said.

---

* After Schmidt's death in 1933, his disciples cast doubt on the definitiveness of his Sargasso proposition. They showed that he had concealed certain data to make his case more plausible and questioned how he could say with any certainty that this was the only eel breeding ground when (a) he hadn't witnessed an actual hatching and (b) it was virtually the only place he had looked. Such criticism, though compelling, does little to diminish the profound story of eels he conveyed to the world, especially since the southern Sargasso Sea was later confirmed by other researchers to be the spawning area of the Atlantic eels.

"It's an issue of scale—the ocean is *huge*. To get where eels are spawning, it's statistically very low probability. Almost impossible. You'd have to be very lucky." It didn't help that every previous year in which they'd made a cruise to look for the adults they'd run afoul of the elements. "I can't remember a single eel cruise when there hasn't been a typhoon that's caused us to change course," Miller added. "It's almost like Poseidon is trying to keep the eels' secret."

That's the greatest beauty I find in eels: the idea of a creature whose very beginnings can elude humans, and the potential that idea holds for our imaginations.

Like the people I met in my travels, I get a good feeling from eels. The nights and early mornings I've spent with them during the fall migration have pulsed with energy and light. Standing in an eel fisherman's river weir in the cool September dark, watching the vein-like ropes of fish fill his womb of wood and stone, I've come to believe the Maori yarns about encounters they've had with the water guardians.

We allow ourselves to believe that nature can be explained. In the process we confine nature to those explanations. The eels, through their simplicity of form, their preference for darkness, and their grace of movement in the opposite direction of every other fish, have helped me to see things for which there is no easy classification, things that can't be quantified or solved, and get to the essence of experience. They have been my way back.

chapter two

TO THE *Sargasso*

The eel weir below Peas Eddy

A November day, driving near the Delaware River in the Catskill Mountains of New York State, I followed a sign and a dirt road it pointed down. Nailed to a post in the tall grass and weeds, the sign said "Smoked Eel" and led me past the Cobleskill stone quarry and a transfer station to a deeply shaded and damp hemlock glen dotted with half-naked hardwood trees. The nearest village, Hancock, on the Pennsylvania border, was run-down, and its residents seemed either to have been cast out there by accident or to have been born there and stayed. Who, I wondered, had cast themselves yet further out to live at the end of this road, and what did smoked eel have to do with it?

At every turn, when I thought I was lost, another sign nailed to a tree confirmed my way: "Smokehouse," "*Casa di Fumo*," "Eel." The road became narrower and more pitted, and just as I was becoming sure that this wasn't the way, I came to a small tar-paper shack with a silver smokestack, perched on a high bank overlooking the East Branch of the Delaware River. A man with a pointy white beard and a ponytail who resembled a wood imp hopped from the plywood door of the smokehouse. His name was Ray Turner.

A gray sky threatened snow; leaves fell and faded on the muddy ground. Highway 17 and the plaintive whine of tractor-trailers seemed far away. Ray, wearing an apron over his shirt, shook my hand like he'd been expecting me. He led me into the smokehouse. There was a refrigerated counter full of smoked foods for sale—trout, salmon, Cornish hens, and shrimp, but mostly eels that he had trapped in the river. And on the adjacent wall were photos that showed how he caught them.

Every summer when the river is low, Ray refurbishes the V-shaped stone walls of the weir that funnels water through a wooden rack designed to trap fish. It takes him the better part of four months to complete the walls, each about three hundred feet long, in preparation for the eel run that occurs during just two nights in September. The run corresponds with the new moon and floods brought on by the hurricane season, when the sky is at its darkest and the river at its highest.

In the smokehouse, Ray took out his detailed record books, conjuring the seasons he spent at Green Flats, by reading from the pages—swarms of yellow butterflies on a dead shad, storms and flood levels on the river. There were lists of things he'd caught in his trap: eels, shad, striped bass, smallmouth bass, brown trout, Ivory soap—"and it was Ivory!"—a rat, a water snake. He described the run of eels in almost biblical terms: his weir receiving the manna of the river upstream. Large numbers of eels overflowed the weir and trap, many making their way around, over, and through, headed to the sea, where they would spawn and die.

"I have three jobs," Ray said, "this smokehouse, gutting the fish, and that stone thing out there." Ray led me down a hill behind the smokehouse to the riverbank and pulled a canoe out of the tall grass. He instructed me to do the same.

"Those flatlanders," Ray said in his direct, military way of speaking, "summer people, yuppies, come up and buy a place here and get a canoe, tie it to the bank. They don't understand the concept of a flood. You know what happens when their canoes tear free? That's right—they float down and end up here. I've got six or seven. Never bought a canoe in my whole life."

We paddled, each in our own canoe, from Green Flats upstream to the weir. Ray kneeled in his canoe, staying tight to the near bank. With his long beard, the hills of the Catskills and the rusty yellow foliage of beech trees behind him, he looked like an old Russian bush guide making his way up the Amur. Truly, he was an anachronism here, a two-and-a-half-hour drive from Manhattan.

In this broad valley, reminiscent of a nineteenth-century painting of the Hudson River school, the weir made an impressive piece of land art. It seemed to change the landscape itself, make the mountains more majestic and the sky bigger. Ray spoke of it in metaphorical terms. "This is the womb," he said as we tied up the canoes and stood perched on the rack. "Those are the legs." He gestured toward the stone breakwaters stretching diagonally on either side of the river. "You see? It's a woman. All the river's life comes here."

I looked at the stone walls of the weir, built every year by hand, that coax the river into the trap. Ray was careful to explain that the weir is not a dam—water percolates through it, over it, and around it.

The trap itself, or rack, looked like a hermit's shack that washed down the river and got stuck here, made of plywood and two-by-fours, erected and taken apart every year. Like a polite host embarrassed by an untidy home, Ray explained that this weir was two months past being in tip-top shape and the wooden rack was about to be disassembled, the parts stored in his basement. As the river moved through the slats, the rack collected all kinds of debris, leaves and branches that Ray removed with a rake as we talked. He explained how the eels get pinned underneath the overlapping ramps, unable to fight the force of the water coming over the top of them.

Ray shared the physics and hydrology of his weir and trap, putting to use his training as an engineer in the army. "There are two wings of stone, as you can see." We waded into the water above the weir. On the inside of the V, the walls were steeper and "chinked" with lots of small stones and quarter-sized gravel. On the outside of the weir the walls sloped more gradually and were covered with beautiful flat paving stones that fit together like puzzle pieces. Ray was intimate with every large stone; some were so heavy, he admitted, he needed the help of three or four friends to move them. A few of the biggest ones were granite sidewalk slabs from the village of Hancock, dragged there by horses on a stone boat in the 1920s. A large rectangular stone that Ray moves back into place every year will be his tombstone when he dies.

"All summer in the low water I'm working on the weir," Ray said. "I get some help from a fifteen-year-old named Jaime and a few friends, but other than that it's all me."

All of this was in preparation for a handful of nights when Ray's yearly catch was made. "How do you know when the eels are going to run?" I asked Ray.

"I say what my brother used to say when he was a bear guide in Alaska—look for signs." A few days before the run, a number of large eels show up in the trap, one here, one there—then the next day ten, then forty, then a hundred. Ray calls these early fish the vanguard eels. The storm that triggers the run could be a thousand miles away, he says—a hurricane may only just be hitting the Gulf coast and it's sunny and clear on the Delaware River—but the eels know it's coming. The eagles show up and get agitated. It's close to the new moon. Days later the rain comes and the river level rises. The water gets off-color.

And if all this happens within two weeks either side of September 27, the eels run. Everything has to be ready. He can get over a thousand eels a night—a ton in two days, a hundred an hour. He fills the canoe with eels and brings it down to Green Flats, dumps the eels in the tank by the smokehouse, paddles back up to the weir, fills it with eels, brings it down again. "All night I'm going," Ray said.

Ray is one of just a few individuals on the entire Delaware River system who is permitted to operate an eel weir. He inherited the right to build it from his father, Ray senior, who had inherited it from his predecessor, Charlie Howard. You can't really say he inherited the weir itself because, like every human structure, it is to a certain extent fugitive; in fact, it was abandoned for a number of years, until Ray revived it. Every year he repositions the stones into walls, using manpower and ingenuity, and every winter the ice and floods take them down again.

"I believe that work is a privilege of life," Ray says.

On old topographical maps, the location of Turner's eel trap is marked Eel Weir Hollow. There has been a weir there on the East Branch of the Delaware River for at least a century, and it is likely that Native Americans had an eel weir at the same site, or at a site nearby, for thousands of years before. Charlie Howard operated the weir in the early part of the twentieth century. When he died, in 1948, the front-page headline in the *Walton Register* read, "Hancock Recluse Lived in Lonely River Cabin."

The obituary, which Ray later showed me a copy of, described Howard as being "as close to the popular idea of a hermit as was

Thoreau," who "conducted an eel rack near his home, gaining part of his sustenance from the eels." Howard was an eccentric whose personal belongings included a box of discarded women's shoes and a jar of silver dollars. He lived upstream of the town of Hancock, in a tar-paper shack by the river. Ray Turner now lives across the river from that old residence and may be Charlie Howard incarnate.

Back atop the rack, we stood and stared up the river at the V-formation of flat water ahead of the walls. A braid of water about six feet wide slipped into the rack like a lacy veil, then splintered through the wooden slats, forming a frothy water-fall beneath our feet. I stared upstream at the force of all this water coming toward us and I got scared, a genuine feeling of fear.

"The river is a formidable thing to tame," I said to Ray.

"We're not here to stop the river," Ray said, "we're here to catch eels. That's our creed. If you're trying to stop the river, it just shows how fucking stupid you are."

On the hill above the river was a small house that looked abandoned. That was roughly the site, Ray said, of Charlie Howard's shack.

"Somewhere I have a picture of him," Ray said, "all grainy and fuzzy, but you can make it out, him on the weir with a net. He died sixteen days after I was born."

From the top of the rack we had a view of the changing light over the river and hills. The trees and the Japanese knotweed on the sandy riverbanks had lost their leaves. Ray pointed overhead.

"There's Baldy," he said.

As I looked up I saw a mature bald eagle fly low over us and land in a tall tree on the far side. Ray began again to rake leaves that had got caught in the slats of the rack.

"When I'm catching eels there's four or five of them here," he said of the eagles. "I see them every day—it's nothing for me."

"Wow!" I said, in awe of the huge bird and struck by Ray's apathy. "You don't have to go to Alaska like your brother to see the wilderness, do you? It's all here. Didn't you say your brother is up in Alaska?"

"He's dead," Ray said. He kept raking. "Actually, he was murdered."

"Geez," I said. "I'm sorry. Up in Alaska?"

"No, here—down the drive where the paved road meets the dirt road. He and some buddies got in a fight and were outnumbered. My brother was killed."

The water gurgled nervously under our feet and the eagle kept his perch on a tree downstream.

I watched the river for a while and Ray put down his rake. We turned around to look upstream. Ray took a deep breath and exhaled. I sensed by the way he was breathing that I should prepare myself for a little philosophy.

"Up here," he said, "it's me and God, and that woman's singing." He gestured toward the river, the "woman" he was speaking of. "It sings to you with a splash and bubble. When the flatlanders have gone back south and the eels are migrating, most of the time when I'm harvesting at night there are no lights anywhere—but the Big Dipper, on the night of the run, is framed by the hills there, just downstream. I love building the walls. The harvest is hell."

⁓

Back at the smokehouse, Ray showed me the two concrete-block chambers where the eels—dressed and brined in salt, brown sugar, and local honey—are hung on rods. Behind each smoking chamber is a fifty-five-gallon drum stove with a door on the front and a chimney hole with two pipes in the back. Once the fire is going in the stove, Ray directs the heat and smoke into the chamber, and the eels are cooked at 160–180°F for a minimum of forty-five minutes.

When the September run is good, Ray can take up to 2,500 eels. This year he had taken exactly 2,406 at an average of .85 pound per eel—just over one ton. Ray kills the eels by putting them in buckets of salt, turns them in a cement mixer full of gravel to take the slime off, then guts them with a knife, cleaning the fish's cavity with a stainless-steel spoon. The hot-smoked eels are sold to passersby, as well as to restaurants and a few retailers. Ray keeps some for his own eating. "Always check the product," he declared. "I consider the eels to be the best-quality protein in my line—a very unique flavor of fish, applewood smoke, and a momentary lingering of dark fall honey. All the fish I smoke, trout and salmon, are farm raised except the eels. The eels are wild; they're like free range."

Ray ushered me through the back door of the smokehouse, past neat stacks of hand-split applewood, to a large wooden tank, like a giant wine cask cut in half, covered in moss and dripping water through its swollen slats. I peered over the chicken wire around the rim into a clear, dark pool. Ray stirred the water with a net, agitating some five hundred dark silver eels, most about as big around as silver dollars and up to three feet long. They were lithe and sensual in their movements—just magical.

"Every year," Ray said, staring into the black water, "I let the biggest girl go. When I finish this year, she's free." He pointed. "Nearly five pounds."

Ray has his opinions on where the eels go. He agrees with the prevailing theory that all the freshwater eels on the east coast of North America spawn in the Sargasso Sea. There, eels that grew up in rivers from Florida to Maine and beyond all meet in the same general vicinity, mixing their collective eggs and sperm in the warm currents in an orgy referred to by scientists as a panmixia. But Ray doesn't buy into the popular idea that eels die when they spawn. Based on the length of the arduous journey and the fact that no one has ever seen an adult eel returning to freshwater, it is a viable theory. Still, this has never been proven very directly.

"No one knows," Ray said. *"No one knows."*

He and I clung to the tank edge, mesmerized by the dark fish writhing at the bottom. I wanted to glimpse Ray's world again—I wanted to see the run.

At the end of the day we retired to the warmth of Ray's home. Snow began to fall outside.

"First snow of the year," he said, petting his dog, named Smell. "And yet I don't like these gloomy days."

He contemplated his work while making me a hot choco-late. Ray seemed keenly aware of the ephemeral nature of our lives on earth—of the rigor of building a wall, watching it get knocked down, and building it again. "It's amazing that a man would do work," he said, "knowing that his work is never com-plete."

Ray grew up in Hancock with his parents and twin brother, went off to college to study engineering, and enlisted in the army during the Vietnam War (he ended up in Panama). After being honorably discharged, he moved back to New York and worked several civilian jobs, from construction to designing and running wastewater treatment plants.

When his brother died, Ray decided to build a house on family property outside of town on the river and live there the rest of his life, reviving the fallow eel weir downstream of Peas Eddy on the Delaware River.

His home is built around a massive chimney that starts in the basement and grows like a giant mushroom through the house. He said that he'd laid every stone in the chimney and every stone had a story. One came from the bottom of the wastewater treatment plant where Ray used to work; another was a geode from Brazil. Some are from the Rogue River in Oregon and the beach in Plymouth, Massachusetts, but mostly they're from his property on the Delaware.

"I live out here 'cause I want to," he said, "at the end of the fucking road." He sat down and rolled a cigarette, using a book to support his hands. "I've seen people come to the head of the driveway, take one look, and turn around. I know what they're thinking—that they just stepped onto the set of *Deliverance*. Those that are brave enough to step out, I tell them, 'This isn't just a smokehouse, it's an adventure.' Most people come, they see the smokehouse, they see the eel tank. You're seeing it all. For you, it's a journey."

chapter three

EELS IN *Maoridom*

New Zealand

Stella August had drafted an itinerary for my monthlong visit, titled "Eel Adventure in New Zealand," and seemed as enthusiastic as I was to learn more from her people about eel traditions. And that's why I believed she would be there to meet me on the day and the hour we had planned.

Eels

Having flown halfway around the world, I was relieved to find her at the Burger King parking lot past the second round-about in Hamilton. Stella stood coolly outside the driver's-side door of her car, a medium-sized attractive girl with long, dark hair, wearing board shorts and a T-shirt with surfer logos.

"Hey there," she said when I pulled up. "Welcome to New Zealand."

I had been working on my book about eels for well over a year when my friend David Seidler, a screenwriter in Santa Monica, asked me one day if I had heard about the sacred eels in New Zealand. I hadn't. David had lived in New Zealand through the 1970s, was married to a Maori woman, and learned about Polynesian culture through her network of Maori family and friends. David told his "mate," a Maori bush guide named DJ, that he had an American friend who wanted to come down to En-Zed to do research on eels in Maori culture. DJ's girlfriend, Nikki, said that her young cousin Stella had recently completed her graduate thesis on the migration of glass eels—that is, baby eels (called glass eels because they're still transparent when first entering freshwater from the sea)—and that she might be just the one to help.

I began an e-mail correspondence with Stella, and she agreed to be my guide. My visit would be an opportunity to listen to men who had grown up fishing and hunting in New Zealand, speaking a language and practicing traditions that had gone largely silent and underground since the arrival of the British in the late eighteenth century.

Stella was just winding down her years as a student at Waikato University and lived, for the time being, near the campus. She shared a flat with her sister, Wikitoria (Wiki), and their best friend, Kare, a law student. There was a general feeling of excitement among the young women because Stella and Wiki had just completed and handed in their master's theses.

"Did you know," Wiki said when I first walked into their place, "this is literally the first day of sunshine we've had in weeks?"

"Yes," added Kare, "we've had terrible flooding, some of the worst in a lifetime. Entire houses have slipped into rivers. Hillsides have collapsed, they're so saturated with moisture, burying roads and sheep and trees."

Stella pointed out that the excess water from the storms, a disaster and annoyance to humans, was an opportunity for eels to escape from inland ponds they've occupied for decades, making a break for the sea, to the spawning grounds.

"I didn't always like eels," Stella said, cicadas humming loudly in the rimu trees outside. "When my father brought them home I wouldn't go near them. I thought they were the most hideous fish ever! I came to love them, but only after spending time with them. They're so cool!" Stella was half sitting, half lying on the couch, backlit by the sun. She and Kare were twenty-four years old and Wiki was two years younger.

"Cool, are they?" Kare said, laughing. "I always thought they were a little strange."

"A lot of people we're going to visit," Stella said, "when I rang them up I told them I wanted to bring an American by who is doing research on eels in Maori culture, the first thing they said was, 'Why does he want to know about eels?' They were a bit suspicious. I told them you had gone all over the world studying eels—then, you know, they started to get it."

"But why would they be suspicious?" I asked.

"Well, it's like this," Stella said. "They've got a lifetime of experience which traditionally they shared only with members of their *hapu*, or subtribe. Why should they share their knowledge with someone who walks in off the street? They're distrustful of science."

"I'm not a scientist," I protested.

"It's not your profession directly, but you operate in part with a Western science mind. They see things differently. To give you an example, I went to this eel conference in Christchurch, where all the leading eel experts in New Zealand gathered. Don Jellyman, probably the most famous ichthyologist in New Zealand—he's at NIWA [National Institute of Water and Atmospheric Research]—was delivering a paper on their attempts to track large migrant eels from the river mouth to their spawning grounds with tags. He explained that the ten satellite pop-up tags attached to ten large female eels had provided very little clear data about where the spawning area was located.[*] When Don sat down, Kelly Davis, whom we will see on this trip, got up to represent the Maori point of view. He addressed Don directly, in front of everyone. 'Our ancestors have known for thousands

---

[*] Useful data on depth distribution and swimming direction were obtained, though.

of years that the glass eels come up the river in spring, and the adults migrate out in the fall. Why do you need to know where they go? What good will it do the fish to find the house where they breed?' Of course, there was very little Don could say."

Not a moment seemed to go by that Stella did not remind me that we would be meeting with the most knowledgeable people on eels in all New Zealand. "Well, as a *pakeha* [foreigner or white person], you wouldn't ordinarily get to see this side of Maori culture," she said to me as we went out to get some lunch.

Over a meat pie and soda, Stella picked up on what Kelly Davis had said at the eel conference. "Why do we have to understand everything that isn't understandable?" she said. "Everyone wants to unlock everything. I'm conflicted because I'm Maori. I don't want to know where the eels go, yet I've studied their movements in the rivers."

The paper that Stella had just finished and handed in for her master's degree, titled "Arrival Patterns and Environmental Cues Associated with the Upstream Migration of Glass Eels into the Tukituki River, Hawke's Bay," was a valuable contribution to the literature of eel biology. And yet the graphs and charts, the seeming absoluteness of the scientific effort, seemed somehow disingenuous to her.

When we returned to Stella's flat, we settled again in the living room. I looked at a few photos in frames on the mantel before sitting down. Wiki, who was reading, folded her book and put it on her lap.

"That's our father, Robert August," she said, pointing to a

photo of a Maori man superimposed over a picture of a misty beach. "His nickname was Falla. He died in a fishing accident."

Both girls had inherited their father's Polynesian features— dark skin and eyes, black hair. Stella later took out an album and showed me a photo of her father hoisting a huge eel speared through the head. He held the end of the spear like a vaulting pole. The eel had its body wrapped around the handle like the snake on the rod of Asclepius, the common symbol of modern medicine.

The subject of conversation drifted to different subjects as the afternoon drew long shadows on the ground, and we cracked a few Tui beers. Talk flowed to things intended to prepare me for what I might see and hear.

The first element of my education was the *taniwha* (pronounced "tanifa").*

"A *taniwha*," Wiki said, "is something that makes itself known at certain times to certain people. Sometimes to warn them of danger, like a guardian. Friends of mine who live on a farm not far from here, at one time or other, have seen this creature with cow legs, half human, crossing their land."

Stella pointed out that the *taniwha* can assume many shapes, but more often than not it takes the form of a giant eel.**

"If you harm a *taniwha*," Stella said, "if you spear or capture an eel that is a *taniwha*, it will cry like a baby or bark like a dog,

---

* In Maori, "wh" is always pronounced "f."

** In the *Reed Dictionary of Modern Maori*, the translation of *taniwha* is "water monster, powerful person, ogre."

or change colors. Something about it will seem strange. It will indicate that it is not like the others. If you kill a *taniwha* eel, you have a *makutu*, a curse, put on you. You start going crazy, like you're possessed. Then you've broken *tapu*—something sacred or off-limits."*

"Spirits usually come out at night," Wiki said, elaborating on the idea of a curse. "Young women are taught not to cut their hair or fingernails after dark, or a spirit can take them and you could be given a *makutu*. There are things like this that our father taught us to do, but we never questioned why. We didn't know they were Maori rituals."

The sisters' Maori learning was strong in areas such as fishing and gathering seafood, but less so in female matters, as their mother was English. At their father's funeral, a family member grieved separately from the rest of the group. Wiki learned it was because the woman was pregnant, and that pregnant or menstruating women were not allowed inside the cemetery—it was *tapu*, forbidden. She wondered how many *tapu* she had broken without knowing it, and how many concerned friends and family had politely looked the other way.

This revelation is outlined in detail in Wiki's own largely autobiographical master's thesis, "The Maori Female: Her Body, Spirituality, Sacredness, and Mana, a Space Within Spaces." It is less scientific and more cultural than her sister's thesis on eels.

---

* The modern word *taboo*, used in English, is a bastardization of the Pacific Island word *tapu*.

A central argument of Wiki's paper is that British colonization upset a balance between Maori men and women and compromised the *mana*, or integrity, of Maori women.

"Balance," Wiki writes, "is an important part of the holistic worldview Maori hold." To the Maori, nature and culture are one and the same.

The Maori worldview acknowledges the interconnectedness of all living things. The colonials—whom the Maori called *pakeha*, meaning "touch of a different breath"—tended to dismantle nature, categorize and classify it. In their lust to find order in nature, to name and possess it, they for the most part ignored the nuances of Maori culture, and ultimately were successful in their mission, whether conscious or subconscious, to subvert it. Partly because of New Zealand's distance from England and the fierceness of the Maori warriors, the British were never able to physically conquer the Maori. But they succeeded in breaking them culturally and spiritually.

In 1840, the Treaty of Waitangi was signed. The treaty was written in Maori and English, and its tenets are vague at best. Maori rights to land and water are contested to this day in New Zealand—the wounds of colonization are still wide open. I could feel the tension in the room with Stella and Wiki, trying to find their own balance between needing to know their Maori father's past and acknowledging that their mother was English, between science and customary beliefs. And though I had met them only hours before, I could already see that the Maori side of them was winning.

"At the end of the trip we'll be staying with DJ at Double

Crossing," Stella said. "That'll be a good time to sit down and make sure we both know where we're coming from." She paused, then said, "About what you're going to write."

Were there things that could not be written about, I wondered, because they were sacred, or because they were incapable of being explained?

⌒

We had a second Tui beer as Wiki prepared a chicken dinner for us. It was now dark outside and a cool breeze passed through the house. I was tired and disoriented from travel. We talked about music, and things twentysomethings talk about. But then an article in that day's paper brought the conversation around to the *taniwha* again.

According to Stella, the Mercer-Longswamp *"taniwha* site" controversy had been in the news a lot of late. It had started two years before when the New Zealand government began to improve the expressway from Auckland to Hamilton. At the village of Meremere there had always been a curve in the road to avoid a swamp—an oxbow of the Waikato River. Transit New Zealand wanted to take the curve out of the road and put the highway straight through. Working in the swamp one day, a member of the road crew lifted up a giant white eel in the bucket of the bulldozer. Many of the construction workers were Maori, and when they saw the big eel, they fled the site.

Since the incident, the government and Transit New Zealand seemed reluctant to legally acknowledge the presence of a mythological creature. Transit New Zealand wanted to straighten the highway across the swamp because they claimed the curve in the road had caused many accidents. The local Maori *hapu*, or subtribe, Ngati Naho, countered that many more tragic accidents had occurred at the spot since construction began, because the *taniwha* had been disturbed and was angry.

Not all Maori agreed. A local *kaumatua*, or elder, Tui Adams, whose grandson had died in a car accident at the site, was quoted in the *Dominion Post* as saying that he didn't believe the giant eel was responsible for the tragedies: "I don't subscribe to the idea that these *taniwha* come out and cause problems because a road's going through. *Taniwha* are actually guardians, they have always been more helpful than non-helpful."

Some thought the controversy had more to do with raising the profile of the Maori fight for customary rights against the mostly white government. The giant eel, long-lived, resilient, and rearing its head, had become a kind of symbol of a cultural revival.

A resolution was imminent, the article in that day's paper said. Rima Herbert, a spokesperson for the Ngati Naho, made a public statement after discussions with Transit New Zealand. "This is a significant cultural site for us," he said, "and we have got Transit to agree to a modification to their design, which preserves most of the site. We have asked Transit to ensure that when critical works are taking place, a member of our *hapu* is given the opportunity to be present to ensure that our cultural values are protected."

On their website, Transit New Zealand's regional projects manager, Chris Allen, did not formally acknowledge that a *tani-*

wha existed in the swamp, but explained the need for preservation of the site in environmental terms. "While a lot of the area we are working in is swamp, this section over about 30 meters appears to be a spring which may be significant for a large stand of kahikatea trees, which need to be protected."

As I wiped my plate with a bit of bread at Stella and Wiki's kitchen table, crinkling the newspaper and reading quotes aloud, I was, frankly, in a state of disbelief. When in America would a highway project be halted because of a god eel? I confessed to the sisters that this was my first exposure to Maori culture, that I was starting fresh, and I sensed an awakening in myself.

I told Stella and Wiki that my only previous knowledge of Maori traditions was from a 1929 book by Elsdon Best called *Fishing Methods and Devices of the Maori*. Born in 1856 in Tawa Flats, New Zealand, Best, the son of British immigrants, is often referred to as New Zealand's "foremost ethnographer of Maori society." I had found the book, a good two-thirds of which was devoted to eel fishing, to be rich and informative. But the very mention of the name drew contempt from Stella.

"Elsdon Best," Stella said—in her Kiwi accent it sounded more like *beast*. "He's not looked on that favorably by Maori. He may have recorded some valuable information about fishing nets and weirs and artifacts and things, but for the most part he wasn't very sympathetic to the Maori way of life."

That night in my sleeping bag on the floor of Stella's room, buzzed and jet-lagged, I remembered my first trip to New Zealand seven years earlier—a trout-fishing trip with my best friend after graduating from college. We caught many large brown trout, hiked in the mountains, and camped under brilliant stars. We nearly drove a rental car off a ravine into a river, and in general had a great adventure. Yet for me the trip had been disappointing. After a month of bumming around the North and South Islands, I left New Zealand without ever really feeling any connection to the place. The people we met in bars were transient Kiwis of European origin, or seasonal Brits shearing sheep. If we came across Maori, they were off in a corner, tight-lipped. I couldn't remember having a conversation with a single one. This trip, I felt, drifting off to sleep, would be different.

❧

The first written account of a *taniwha* is most likely the one recorded by Captain Cook on his 1777 voyage to the South Seas. While anchored in Queen Charlotte Sound off the South Island of New Zealand, Cook wrote the following about statements by a local Maori: "We had another piece of intelligence from him, more correctly given, though not confirmed by our own observations, that there are snakes and lizards there of an enormous size. He said, they sometimes seize and devour men."

Maori would not have known a lizard of the size Cook related, and would never have seen a snake, as there are no native snakes in New Zealand. What he described, "eight feet in length, and big round as a man's body," was most likely a longfin eel.

Taniwha or not, the longfin eel is an impressive creature. Like other members of New Zealand's distinct fauna—the now extinct moa birds (killed off by the first Polynesian settlers for food), more than twelve feet tall; the native kakapo (the world's

largest parrot, now endangered); and the largest living insect, a
kind of cricket called a *weta*—the longfin eel has a tendency toward gigantism. Capable of growing to over eighty pounds and
living more than a hundred years, it is the largest and longest-
lived freshwater eel species in the world.*

The longfin has been a consistent and available food source
for Maori as long as they have been on the islands, and for that
reason, along with its impressive stature, it has garnered awe
and respect, inspiring a good share of stories.**

For all his apparent shortcomings, Elsdon Best devoted hundreds of pages of his books to the Maori's long association with
the eel. Through his work we are led to believe that the eel is
among the most important creatures in Maori culture, outshin-
ing the shark, the whale, and the kiwi bird put together. By sheer
mass and availability of protein, the eel was at certain times of
year the Maoris' most important source of food. Best recorded
more than three hundred local Maori words for the freshwater
eel in its different life stages.

I was yet to see my first New Zealand longfin eel, but Stella
assured me that this would soon change. As my guide, she felt

---

* With the exception perhaps of the tropical eel, *Anguilla marmorata*.

** The endemic longfin, *Anguilla dieffenbachii*, and smaller native shortfin eel,
Anguilla australis, which is also found in Australia, are by far the largest of New
Zealand's twenty-five native freshwater species of fish. Of the two eels, the longfin is longer-lived, and can be distinguished from the shortfin by a dorsal fin that
starts closer to the head.

that it was important I experience the subjects of our adventure before we started hearing stories about them.

About ninety miles southwest of Hamilton, in a coastal village called Kawhia, an elderly British woman was feeding eels in a small spring-fed stream in her backyard. Such feeding areas were relatively common in New Zealand but were normally kept quiet to avoid poaching of the valuable eel meat.

Traditionally, Maori had sacred eel ponds, where the eels would be fed daily. Sometimes the eels were brought to these ponds and kept there, with no access to the sea, where they would live, some Maori say, for hundreds of years. But most of the time eels were kept in the pool of a creek or river and could come and go as they pleased. Often they stayed for a long time, perhaps because of free meals or the love that was imparted by the people around them. But the eels' ultimate goal was to store up energy to get to the spawning grounds. No one had yet found the spawning place of the New Zealand longfin eel, though scientists suspected that they migrated to the north and spawned near Tonga over the Kermadec Trench.

"The eel's life is about travel," Stella said to me as she drove. "The only reason they stay in the stream is to store food for their long journey." The longfin eel travels from the sea to the headwaters of rivers and streams, spending an average of thirty years in freshwater before returning to the sea to spawn.*

---

* The shortfin eel does not go up rivers as far as the longfin, spending its life in estuaries and lowland lakes and rivers.

"The eels' movement is universal in Maori culture," Stella said. "As the eels move, they leave the path of life behind them."

We exited the main highway and drove through rugged farmland to the coast. Intermittent lush green stands of native bush—*ponga* (tree ferns) and tall grasses—held on to the slopes of steep ravines, land that could not be farmed. Stella pulled the car off at a convenience store, and I followed her to a back shelf, where she grabbed a few cans of dog food. "For the eels," she said.

"How big are these eels?" I asked Stella as she brought the cans to the counter. "I mean, are they really as long as a person?"

"You'll see," she said.

Down the road a bit further the smells of the sea became stronger. Stella slowed the car when we were in full view of the ocean. She pulled off the road at a moss-covered gate where a small sign hung; on it was painted a black eel. Stella opened the gate and we drove down the gravel drive to a modest ranch-style home.

Barely visible beside the house was a pool of water that collected seepage from a cold spring. Below the pool ran a small brook, choked with watercress. Stella and I got out the cans of dog food and opened them. An old man came out of the house to greet us.

"We've come to see your eels," Stella said.

"Oh, them eels isn't mine, they're her thing," he said, referring to his wife. Soon the wife appeared and the old man went inside. She introduced herself as Beryl and described how the pool of eels had come to be.

"We moved here ten years ago. The place was covered in blackberry and gorse. When we cleared the yard we found the spring, and dug it out just enough to hold a few ducks."

Soon the crystal pool was rimmed with a green collar of watercress, and large dark shapes began to appear as if from nowhere, attracted by the bread that Beryl fed her ducks.

"I don't know where they'd come from. I guess they'd always been there." She looked into the pool. "They just keep getting bigger and bigger. They're my babies, just like the three bulls, the two goats, the dog, the dozen odd chickens."

The ducks, Beryl added, had relocated to another pond on the property since the eels had showed up.

"Are you ready to see some eels?" Stella asked.

"Let me get a bit of meat," the woman said.

"Oh, don't go to the trouble," Stella said. "We have dog food."

The old woman insisted it was time for her to feed them anyway, and soon returned from the house with a bit of steak. She tied the steak to a string and we watched her wade out into the shallow pool in her gumboots. As she waved the steak in the current, I saw a few large heads emerge from the watercress, seemingly from nowhere. Giving in to a natural reflex, I took a step back.

"Don't be afraid," Beryl said, "they won't harm you. Unless you're holding food—they might bite you by accident."

As Beryl lifted the steak on the line out of the water, a huge eel, about as big around as the calf of her leg, lifted its head out, dancing to and fro to keep its body up, not unlike a cobra.

"Oh my God!" I said aloud, my mouth agape.

When she lowered the meat into the water, five or six big eels, their heads five to eight inches across the back, vied for a piece. They grabbed on, making loud sucking sounds to try to get an advantage on the steak, and then they rolled their bodies to tear pieces off.*

Stella had taken off her flip-flops and was walking bare-foot across the grass. She spread the contents of one of the cans of dog food on the grass near the edge of a pool. With a stick, she pushed some of the meat chunks toward the pool. A single big eel came to the rim of the concrete ledge to investigate. It sniffed a few times, then tilted its head and body, propelled itself over the ledge onto the grass, and began taking pieces of the dog food into the side of its mouth. A few smaller eels followed, and soon the grass was wet with slime.

I could see their features clearly: wide mouths, broad lips, and nostrils like tubular horns. These eels were big, but the biggest eels in the pool barely approached the ledge, hanging back in the darker depths. Once in a while I caught a glimpse of a real monster, exposing its head from the thick mats of watercress, but never its whole body.

---

* An eel can rotate fourteen times per second.

Stella feeding eels by the pool at Kawhia

Stella squatted on the grass, her long black hair nearly touching the ground, letting the eels glide up between her bare feet, touching one and then another on the top of the head, petting them.

Tuna—the Maori word for eel, and on some Pacific islands a synonym for the phallus—is also the name of a prominent figure in Maori mythology, a god in eel form, often found wrestling with Maui, the Pacific islands' equivalent of Hercules. In one story, which varies from place to place, Maui finds Tuna in bed with his wife, Hine, while she is sleeping. Maui chops Tuna in half, the head becoming all the freshwater eels of the world and the tail all the saltwater eels of the world.

It is thought that the stories in which the snake plays the role of monster-seducer and/or guardian trickled down into the Pacific from India and Indonesia to the islands of Micronesia and Polynesia. The Pacific Islands, however, are largely devoid of native snakes, so the role is assumed by the creature closest in appearance and movement, the eel.

The eel is not always simply an unwelcome seducer in native stories; sometimes it is a pet or a lover. One of the most common eel stories in Polynesia involves Tuna the eel and a beautiful girl named Sina.

One day Sina goes to the spring-fed pool to get water for her mother's cooking. She dips the pot in the water, and when she gets home, she realizes there's a baby eel in it. Sina nurtures the eel as a pet and comes to love it. When it grows too large for the pot, Sina's mother puts it in a tub of water outside their home. The eel gets bigger and bigger, outgrowing the tub, so they release it back in the spring hole it came from. Every time Sina goes to bathe in the pool the eel comes out and swims around her playfully, but eventually it gets so big that Sina becomes afraid of it. One day, Sina is doing her washing in the pool and the eel wraps around her leg, violating her with his tail. She runs home to the village and tells her mother and father, and a warrior from the village goes down to the spring to kill Tuna the eel. Sina follows, and as the warrior is about to chop Tuna's head off, the eel asks Sina to bury his head in the sand and then visit the site day after day. Sina promises she will, and cries because she still has love for Tuna, who had once been her pet. When the warrior kills Tuna, she takes the eel's severed head and buries it in the sand, and from that spot grows the first coconut tree. The Samoans and other islanders say the eyes of the coconut are the eyes of the eel, and the soft part below the eyes, from which you drink the coconut milk, is Tuna's mouth. So every time Sina drank from the coconut, she was kissing Tuna.[*]

Tuna's head and the coconut tree

What I'd witnessed on the banks of the spring-fed pool in Kawhia was some strange reenactment of a Pacific Island

---

[*] The white meat of the coconut is sometimes called *te-roro-o-tuna*, "the brains of Tuna." In some versions of the story, Sina, also called Ina, falls in love with the eel, named Tuna, who can take the form of a handsome man. One night there is a big flood and Tuna, in eel form, offers his sacrifice in order to stop the flood.

In his book *The Masks of God: Primitive Mythology*, Joseph Campbell juxtaposes the Pacific island story involving the eel, the woman, and the food tree, with the biblical story of Eve, the snake-seducer, and the fruit in the Garden of Eden:

myth—at least in terms of the girl and her affection for the eel. The vision of Stella by the spring, with the eels at her feet, compounded the awe of seeing such a large fish feeding out of water. But somehow, too, it breathed life into what I had felt were beautiful but otherwise lifeless stories. Seeing the large eels and Stella together, I instantly understood that what I had been getting in books was a very small part of a deep and old relationship that carried the weight of time. Expecting to understand the impact of a Polynesian myth from just reading it was like trying to know the glow of a flowering plant from its pressed and dried blooms. The oral stories had been compromised the minute they were written down, then more so when they were taken out of the environment in which they were created.

I only later realized, during further travels in Polynesia and Micronesia, why the stories of Hine, Maui, Tuna, and the *taniwha* had seemed pale to me on paper. They evolved to be heard, and not just anywhere, but amidst the sublimity of nature: in a dark forest, near a booming waterfall. The tales did not transcend their original contexts well.

From a Darwinian point of view, that was the error of the

"Paradoxically, then, it would appear that although we are moving eastward into the Pacific we are also coming closer to the biblical version of the mythological event through which death came into the world; and something rather startling is beginning to appear, furthermore, concerning the relationship of Mother Eve to the serpent, and of the serpent to the food tree in the Garden. The voluptuous atmosphere of the lush Polynesian adventure will be different, indeed, from the grim holiness of the rabbinical Torah; nevertheless, we are certainly in the same old book—of which, so to say, all the earliest editions have been lost."

indigenous people's spiritual platform. The survival of the Maori faith (as with the faiths of other indigenous peoples with animal deities) depended on a connection with nature being maintained, and also on nature remaining intact with all its creatures. When colonists came from Britain and elsewhere and developed the land, the wilderness became fragmented, and so did the native people. Movies about modern Maori, such as *Once Were Warriors* or *Whale Rider*, show a broken people struggling to keep a foothold in cultural forms that existed before colonization, while having to adapt to the patriarchal structure and hierarchy of the Western world. Because the health of Maori culture relies on the health of nature, the Maori cultural resurgence, which Stella and Wiki are at the forefront of, is inevitably an environmental conservation movement. To resurrect a nature-based spiritual society, you need nature to be intact, to protect the sources of awe that inspired spirituality in the first place. What happens to the *taniwha* if the giant eels that brought that monster to our imaginations become endangered or extinct?

I had been awakened by what I had seen: the longfin eel, a magnificent creature, a living myth. The mass and muscle of the large eels had illuminated the Polynesian stories I'd read at home. I understood the impossibility of a foreigner like me ever fully grasping the nuances of the Maori spiritual world, but having grown up in nature and been moved by its wonders, I felt I had an inkling of that connection with what has been referred to as the numinous.*

In reading Elsdon Best, I had gotten the sense that he understood the limitations of his recorded material regarding Maori culture. He had felt the gravity of the stories the Maori told him, but knew he was not getting the whole picture—and even if he had been, he felt, he would never be able to fully communicate it. To add to his frustrations, the culture he was attempting to record was changing rapidly before his eyes. In one instance, while recounting a tale of Maui and Tuna that a Maori elder had told him, Best wrote: "It is not a good illustration of a Maori recital, approaching as it does too closely the clipped, cramped, unadorned modern style of diction." He was witnessing a profound transition in mind and spirit of the native people, whose culture, language, and religion were being supplanted.

Where the nature-based faith of the Maori was failing, the imported Anglican faith of the British colonists flourished. Christianity is portable. It does not rely on anything being intact. It

---

* Joseph Campbell writes in his 1976 book *The Masks of God: Occidental Mythology*: "In the long view of the history of mankind . . . essential functions of mythology can be discerned. The first and most distinctive—vitalizing all—is that of eliciting and supporting a sense of awe before the mystery of being. Professor Rudolf Otto has termed this recognition of the *numinous* the characteristic mental state of all religions properly so called. It antecedes and defies definition. It is, on the primitive level, demonic dread; on the highest, mystical rapture; and between there are many grades. Defined, it may be talked about and taught; but talk and teaching cannot produce it. Nor can authority enforce it. Only the accident of experience and the sign symbols of a living myth can elicit and support it."

can be practiced and understood in a city or in the countryside, under a roof or out in the open, by any people anywhere. But the faith of the Maori is specific to New Zealand. It cannot be easily packed up and taken somewhere else. Polynesian faith would not be comprehensible in Alaska, where the totems are eagles and bears instead of eels and kiwi birds.

Many of the elder Maori that Stella and I visited believed that colonists ultimately diminished the Maori not with guns but by cutting the forests, building dams, introducing insecticides and herbicides, and making the native bush less and less contiguous. The colonists brought with them their own religion to replace the indigenous one, but they also brought their own familiar species of animals—the trout, the hare, the stag—that in some cases successfully displaced the native creatures, the totems of the Maori.*

The next morning Stella and I set off from Hamilton toward Hawke's Bay and the more remote east coast of the North Island. We were headed to visit our first Maori elder, Brown Wiki, a history teacher at Hastings Boys High School (and the father of Stella's flatmate, Kare).

---

* To be fair, the early Polynesians were also responsible for introducing non-native species. They brought their own creature comforts with them to New Zealand: the dog in particular and (whether intentionally or not) the *kiore*, or Polynesian rat. The eighteenth century saw the introduction of the pig by Captain Cook, leading to the local name for wild pigs, *Captain-Cookers*.

Brown was an imposing man—broad-shouldered, heavyset, with dark freckled skin, big round eyes, a flat nose, and purplish lips that looked like they had been painted. With permission, I set down my small digital recorder and took an account of what he chose to share, making occasional notes in my spiral notebook. He had agreed to our meeting as a favor to his daughter. When he spoke, he mostly addressed Stella.

"Well," Brown asked, "what would you like to talk about?" Stella explained to him why I had come. Outside, schoolboys in uniform played handball against a wall. A mild breeze blew through the classroom where we sat.

Brown said that every *hapu* has its own eel stories and every individual in that *hapu* has his or her own version of those stories; this makes the stories very personal and diverse. Brown said that although Maori culture had been forced underground by colonization and the language forbidden in schools, it had always been there and was coming back with force. Like Stella's father, Brown had learned Maori language at home from his grandparents.

"The older men we'll be visiting," Stella had said, "they grew up speaking Maori in the home. They'll be telling a story to you that they heard in Maori, and as they tell it, they'll be translating it in their heads. Words have different meanings in different sentences. In different circumstances different words apply. When the Maori language is translated into English, it makes for beautiful sentences."

"Eels did not live on earth originally," Brown began in his deep voice with its Kiwi accent. "The eels were all up in the heaven. When the planet they were on came too close to the sun and was too dry, they followed Tawhaki's trail down [Tawhaki is a godlike being connected with thunder and lightning]. Here on earth they found plenty of moisture, plenty of water. The eels we eat today are small edible eels compared to the big ones that followed Tawhaki.

"The Pakipaki [River] down the road had a guardian eel. The eel was guarding the sustenance of the family. The family knew which eels to keep for consumption and which to keep as guardians. They could tell by the shape of the head and the color of the eyes. The really big eels, their eyes can turn red. Maori people treated guardian eels as part of the family, feeding them and bringing them offerings."

Brown proceeded to tell us about how construction on State Highway 1 to straighten the stretch of road through the swamp in Meremere had been stalled by the guardian eel. His story was similar to what the newspaper and Stella had said on the matter, but in his own words.

"Why didn't they take the road straight through in the first place?" Brown asked. "Because, well, there's a good reason— apparently that's where the *kaitiaki* eel, the guardian eel, lives, in the swampy part of the bush.

"And apparently there was a digger, trying to drain the water, came up with a huge eel. It was a *huge* monster. And that's why, going through the bush, they halted. And had to agree to divert the road again.

"But you see, some of these things, they either . . . they've got no understanding of legends like that. They were up against something they ignored. We can laugh, but there is still things like that. What they were up against was real, it's real, those things."

Brown said that the eels in the Meremere swamp came out of the Waikato River, which is known for having some of the largest eels in New Zealand. He said that the Waikato *taniwha* eel is so big that it has to go out to sea to turn around, and the average eels from the river are large enough that you can use their skins as riding chaps.

In freshwater, Brown said, the *taniwha* usually takes the form of an eel. In salt water it is most often a shark. "Where I was brought up the eel was not a staple. We lived off the sea. When we fished for sharks it seemed like you'd smell like shark forever. They would call in whales to the harbor with a conch and do an organized culling of the young. Our people came to New Zealand following the whales from Tonga and Samoa in A.D. 950. They followed the whales along the currents. Some migrated back to Tonga with the whales."

Sometimes the *taniwha* is an outright angry monster; other times it is a kind of guardian of the resource that only acts when humans break *tapu*, some sacred restriction. In Hawke's Bay there is a story of a guardian that took the form of a shark named *moremore*.

"We had a friend who went out to dive for shellfish," Brown went on. "We told him, 'Don't be so greedy, don't take so much.' But one day he was bit by a shark. Then on his way back he was stung by a stingray. The guardian shark warned him. The shark made his presence known."

⁓

Driving away from the boys' school, Stella said to me that Brown had been holding back. "Because you were in the room," she said. "Because he was trying to explain things on your terms." We drove on for some distance before she spoke again. "What you understand as myth," she said to me, "in our culture, they are not myths—they are stories about real events. Before, the mystical was part of everyday life, the real mixed with the fantasy. There was no distinction made between the two."

She continued, sounding slightly irritated. "If I were writing a book," she said, "I would not try to describe what a *taniwha* is. A *taniwha* is completely personal—your opinion of it is based on your own experiences. Yes, it is a mythical being in Maori culture, but there are differences between *hapu* and between *iwi* [tribes] all over New Zealand."

Elsdon Best, in his book *Maori Religion and Mythology, Part II*, wrote of the *taniwha* in the past tense. "The taniwha," he writes, "were supposed to dwell in remote places, in the depth of the forests, on rugged mountains and high bush-clad ranges, in broken country where cliffs, canyons and caves are found, and in deep-water lakes, rivers and ponds." He seemed to suggest that the *taniwha* had somehow gone extinct before the end of the nineteenth century. "As our Maori folk become more and more Europeanised," he observed, "one hears less of supernormal beings and miracles." What, I wondered, did Best mean by this? Was he being ironic? That one hears less of supernormal beings because they never existed? Or was it that Maori had chosen to speak less of things such as *taniwha* because they'd learned a foreigner could never grasp the concept of one, and that even if they tried to explain their belief in them, they would only be ridiculed?

Stella did not believe that Best deserved any credit for subtlety. "They were not sensitive," Stella said of the European ethnographers of Maori culture. The authors who recorded the oral stories of native people, she told me, saw them as quaint folktales, because they did not believe them.

"I believe that life is governed by many unseen forces," Stella said emphatically.

As we neared the city of Napier, the presence of the sea was palpable. We drove by vineyards glimpsed through gateways between high hedgerows. This region, along the coast called Hawke's Bay, was in the territory of Stella's *iwi*, Ngati Kahungunu (pronounced "nati kahununu"), and we had an appointment to meet some of the tribal leaders at the *runanga*, or management center for the *iwi*.

"They wanted me to bring you by for an official welcome," Stella said to me.

The *runanga* building itself was modest and unadorned, like a small country school. Stella and I walked into a kind of conference room where I was welcomed warmly by a man named Digger, handsome, middle-aged. A few other Maori men and women sat around a large table. It looked as though we had interrupted a regular meeting. Stella and I sat down in the only two vacant seats.

"Welcome," Digger said, rather formally, "as a visitor to New Zealand, a friend of Stella's, and a friend of our *iwi*. The eel has been considered a pest by *pakeha*, because they prey on the trout, introduced by the British. I hope you will find that the eel is a cultural icon to all Maori throughout New Zealand—rich Maori, middle Maori, and poor Maori. We wish you luck in conveying the importance of the eel worldwide."

Digger then told an eel story. All the others in the room listened.

"I grew up near Lake Waikaremoana," Digger said. "Hundreds of years ago, the lake had been cut off from the sea by a rockslide. No one knows how the eels got there from the sea, though some Maori speculate they've been there since before the rockslide. The eels there are old and big. The females circle around the lake in fall when they get the urge to spawn, trying to find a way out to sea, and they keep circling until the urge goes away. They put their noses up above the surface of the water to try to smell the rain. They're waiting for a big typhoon to come and wash out the rockslide so they can get out to sea. But in the meantime they keep living, and growing bigger. The scientists don't believe that they will keep waiting, that they can live for hundreds of years. I'm sick of scientists saying, 'Where's the proof?'" Digger said. "Saying that is talking against our Maori culture. Eh, Stella?"

Stella sat humbly in her seat and gave a quiet "Yeah."[*]

Stella had used my visit as an opportunity to fulfill an

---

[*] Stella later explained that part of the reason she pursued science was to "provide that proof for her people."

overdue obligation: to give a brief presentation on her master's thesis work, the education that the *iwi* had assisted with scholarships. The *runanga* board members who sat around the table were mostly silent, prepared to listen to what they had invested in.

The *iwi* seemed to inhabit contentious ground, between wanting their young Maori to have the advantages and opportunities (such as a university degree) that would help them compete in a westernized world and being wary and skeptical of that world. Stella said to me later that she had met with a lot of goodwill from her *iwi* for the academic work she'd done.

" 'You're so young, and female, and you've done all this work on eels,' they say. They want to help me. A lot of people have said to me that they wish there were more young Maori people like me doing what I'm doing."

Stella introduced herself: "My mountain is Kahuranaki, my river is Tukituki, my *hapu* is Ngati Kurukuru, my *marae* is Taupunga, my *iwi* is Ngati Kahungunu, and my name is Stella August." And then she read aloud from a two-page abstract of her paper. Her investigations were concerned with how moon phases and water temperature affect the migration of shortfin and longfin glass eels up her tribal river, the Tukituki. Over two different years, September through November (the New Zealand spring), under the stars six nights a week, often with friends, Stella counted the small fish (two to three inches long) that came into her nets. In the first year 50,287 glass eels were captured, counted, and released, and in the second year 19,954. The largest single-night capture occurred in the first year, 18,619 fish, coinciding with the largest spring tide, a new moon, and rising river water temperature (she found that the glass eels were more likely to move upstream into freshwater when the freshwater was warmer than the seawater). Stella explained to her audience that through this work she had formed an even more intimate relationship with the river and the eels.

At the end of the presentation, Digger thanked Stella. The others in the room remained mostly silent and expressionless. They got up and shook my hand and gave Stella hugs as we left.

Everyone we talked to remarked about the big storms that I had just missed. Stella explained that the heavy rains had probably allowed all the migrating silver eels to go downstream, even eels that had been trapped in ponds and puddles with no access to the sea for many years.

"Eels will cross over land in wet weather," Stella said. "On rainy nights on the farm, our cat Buddy brings eels up on the porch with their heads munched off. He catches them in the paddocks. His neck is covered in dried slime where the eel wrapped itself around him."

All along the highway we saw "slips," where hillsides stripped of vegetation by grazing sheep had become oversaturated with moisture and collapsed into rivers and across roads and bridges. Some of the slips looked like avalanches of rock, sod, and mud, maybe with a few sheep mixed in. Riverbanks were cut almost to the steps of homes, roads and bridges were blocked, and some houses had even fallen into the rivers, like one we passed on the Rangitikei River.

⌒

North of Napier at the far eastern end of Hawke's Bay is a settlement called Whakaki, where the people are known as the Eel People. The *kaumatua* we visited at the *marae* (the social and religious gathering place for Maori) in Whakaki was named Walter Wilson.

"We've adapted to the eel title," Walter said. "It gives us a distinction. We're thick-skinned and slippery."

In the town of Whakaki there is a lagoon that is separated from the sea most of the year by a bar of sand and gravel about three hundred feet wide, what Walter called a "shingle bar." Periodically the bar is breached by storms, allowing young eels to enter in spring and mature eels to exit in fall. The unique topographical features of this "lake" attract the eels, and the Maori have been fishing them there for centuries. From the lawn of the *marae* you could see the lake and, immediately beyond, the ocean.

Walter said that traditionally in spring the Maori dig a channel from the sea to provide passage for the glass eels into the lake. In autumn the exodus of adults is more dramatic. The mature eels stage at the ocean side of the lagoon waiting for a storm, and when the big waves are pounding on the beach, they

make a break for the ocean en masse over the shingle bar, sometimes in a huge ball. Walter described this natural phenomenon
as Stella and I sat across a table from him in the *marae*. He rolled
a cigarette from a bag of Port Royal tobacco, lit it, puffed, and
deposited the ashes in a *paua*, or abalone, shell.

"This is when our people fish for the eels, during migration
season, when the eels are staging by the lakeshore. The eels are
waiting for what we call the *tai tipi*, the big tides. I wouldn't call
it a tsunami or whatever, but it's a big wave caused by a storm
out at sea. The eels gather by the thousands, you see them, and
when the *tai tipi* arrives they go, they go, over the land, side by
side, over the sand and the shingle."

I asked Walter what alerts the fishermen—how do they
know when the eels are beginning to gather for the run? He
became slightly defensive.

"I know when to go down!" he said. "But I ain't gonna tell
you. I can't take you to the place we fish. The trustees would
have to decide if that's okay." He paused to tap ashes from his
cigarette. "Why do you want to know? To explain it? The
knowledge is used for money. My opinion is that we give away
too much knowledge."

By "we," he meant the Maori people. He believed, for instance, that the stories told in the recent movie *Whale Rider*—a
modest New Zealand production that had garnered praise from
Hollywood—were used illegally, without permission. "This is
Maori intellectual property," he said.

Walter, who looked to be in his late sixties, had a reddish face
and wore a wide-brimmed hat and a long-sleeved shirt. Stella

and I followed him outside the *marae* into the bright sunlight. He took us in his truck to the beach, a short drive away, to see the dunes and the shingle where the eels cross. It was a black sand beach with pieces of white pumice and driftwood. The surf pounded loudly.

Walter said he couldn't show us specifically where the eel nets were set. Traditionally an elder passes on his knowledge to one person before he dies, and that person isn't searched for but makes him- or herself apparent. The knowledge is not shared with anyone else.

"It's just how it's done," Walter said somewhat apologetically, changing his earlier tone. Then very abruptly, leading us down the sunny beach, he said, "Who owns the sun? Here you are, come see it yourself."

We stood, our clothes blowing like flags behind our bodies. I was looking from the lake to the sea, from the sea to the lake and back, wondering what it would be like to witness thousands of eels crossing the bar.

"The dunes are so dynamic," Stella said, watching the stiff sea breeze blow the sand into the air. "They are always changing."

"That's nature's artistry, I call it," Walter said, lifting a piece of driftwood from the sand.

Noticing that Walter's attitude had softened, I decided to recast a previous question.

"How do you know when the typhoon is coming," I asked him, "the *tai tipi*?"

"You can hear it," he said. "It booms up in the bloody hills.

That's how you know it's a *tai tipi*. The big, big waves come only now and then, and the eels make their way to the crest of the dune, at night, to catch the wave. And then they go. Once they get onto the ridge of the beach and they're heading down into the sea, they're gone."

Walter said that when the eels are ready to leave, they develop a thicker skin, an "extra layer," he called it. Their noses become more pointed, their skulls structurally change, and their eyes enlarge and cloud over in a bluish haze. These physiological changes are consistent with ones that other migratory freshwater eels undergo the world over.

As we were walking back to the *marae*, Walter said to me, "The only reason you're here is because of my connection with Stella's father."

We sat in the *marae*, our faces flush from the wind and sun, and Walter made us a cup of tea. When Walter began speaking, I turned on my small digital recorder and placed it next to my notebook without asking permission. I was afraid that if I asked, Walter would say no.

"That's why I asked Stella how much you want to know. Why do you want to know? For public consumption?" Walter said, addressing me. "I'm not being rude, that's just a thing. This run of eels happened before Maori got here, before humans got here. As long as the lagoon has been here, that's been happening.

"What happens here at Lake Whakaki is unique to this part of the world. The only other place that I know that it happens used to be Lake Ellesmere in the South Island. I've never been there to see it, but they tell me that it happens there, too."*

Walter went on to talk about the health of the fishery and how it's managed. For the Maori, the harvest season is mostly during migration, and if a thousand eels pass through a weir, net, or trap, they take only a hundred, just the ones that look good to eat. The fisheries regulations in New Zealand, Walter said, were faulty, because they managed the fisheries to let the small ones go and keep the big ones. "The problem is that the fish that are released are most often males, because males on average are much smaller, and all the big egg-bearing females are taken. The Crown has the nerve to tell us Maori how to manage a fishery we've been managing just fine for hundreds of years."

Before leaving the *marae*, I asked Walter if he had any thoughts about where the eels go once they leave the lagoon for the sea.

"They're going to die," he said, lighting a cigarette. "They're going out there to die. Once they have laid their eggs and fertil-

---

* The fishery in the lagoon at Whakaki, though unique, is not entirely unlike eel fisheries in other parts of the world where similar conditions exist—where lagoons or lakes with eels are very close to the sea. Comacchio, Italy, in the delta of the Po River near Venice, is a famous example. Lake Forsyth, or Waiwerea, in the South Island of New Zealand is separated from the sea by a gravel bar, shaped by tides and storms, that varies in depth and thickness. The Maori figured out long ago that if they dig channels in the gravel bar from the lake heading toward the ocean, but not entirely through the bar, then during the fall migration the eels will be tricked into thinking there is a way out, and they will swim into the channels, where they are gaffed.

ized those eggs, they all die. And some sharks out there have a great big banquet."

"I'd love to see where they go," I said.

"And I hope you never find out."

As we drove away from the *marae* Stella scolded me. She said that several times during Walter's monologue I had interrupted him to ask him to explain something, or repeat something I didn't understand. Stella was adamant that I not interrupt any more of the people we were going to visit.

"Because Maori is an oral language, you are expected to listen," she said. "That is your role in the conversation. If it's your turn to speak, they'll listen to you. If you miss something that they say, that's your problem. Don't interrupt to ask them to repeat themselves."

⌐

The next morning there was a story in the local paper, *Hawke's Bay Today*, of a *pakeha* boy named Hayden who speared a giant eel in small stream running through his suburban neighborhood in Napier. The piece was headlined "Monster Eel Couldn't Escape Hayden" and showed a photo of the nine-year-old boy triumphantly holding the dead eel draped over his shoulder.

The reporter's slant was that for the boy it was an act of heroism, that this boy had slain a monster. In the days following, letters to the editor weighed in. One reader wrote: "I found the story of the boy and the eel appalling. How would *Hawke's Bay*

Today have reported it if it had been someone's pet rabbit or cat which had been stalked, tortured, stabbed in the head and tail, dragged home, and put in the freezer?"

One Maori man came to the defense of the boy: "I feel it is very inappropriate for mature adults to pillory him and reduce his actions to a shameful and senseless deed. As adults we have gathered a lot of experiential knowledge, and most of us use it wisely. But as a young person, we are still moving into the realm of experience and achievement through adventure."

Later that day we visited a Maori man named Haetia Hihi who had his own take on the story. "It was a fault of the boy's elders," he said, sitting in his home in Napier. "You can't really fault the boy."

Haetia told us a few stories of growing up eeling.

"We'd be sitting with our grandparents by the candle," he said. "When the moths appear that was our signal—while we were having an evening meal. Take a piece of hoop iron and tie it to a piece of wood. You swing the hoop iron over the back of the eel and then gaff it. We call this *ripi*. The water was up to our knees. The *matarau*, the spear, had two prongs made of number eight fencing wire. Some people used a pitchfork. The boys would gaff them. The older people preferred no barb on the spear."

The thing that was destroying the fishery wasn't the boy in the ditch with a spear; it was the issuing of commercial licenses and the catching of eels for overseas markets. Haetia spoke of his people noticing a decline in the eel population. "Lately we put our *hinaki* [nets] in overnight, in our own river, the Esk, but the problem is there are no eels left around."

The absence of eels, Haetia said, was leading to the end of certain Maori rituals. "Eels are eaten at wedding ceremonies. In Runana they still observe these practices of the past to catch the eel for ceremonies, weddings and such. Two or three men get naked and go into a lake. They know the holes in the bank where the big eels live. They get at either end of the hole and one man grabs the eel by the gill and pulls it out. These are big eels, over six feet long." Haetia said at the wedding you see the big eels hanging before they are prepared for the party. But now the *pakeha* farmers have given permission to the licensed commercial fishermen to catch eels there. "They found where their holes were and they took all the big ones."

Stella and I spent the night in a building that used to be a jail, on the property of a couple named Bruce and Kate. Bruce was a fishing and hunting guide on and around the Rangitikei River. Kate's family owned more than ten thousand acres along the river. She was half Maori, though she had fair skin, fair hair, and freckles.

At dinner Bruce told a story of a small pond where they used to shoot ducks.

"The pond was no more than twenty feet across," he said, "a small pond, but ducks came to it and we shot them. There were eels in the pond, lots, and big ones, and they would smell the blood when the ducks landed dead in the water, and if you didn't retrieve the ducks fast, the eels'd grab them and eat them. Finally the farmer who owned the pond decided to take all the eels out because they were pests to the hunters. In that little pond they harvested over three ton of eel, and no one even knew how they got in there."

In New Zealand, it seemed, almost everyone, Maori or *pakeha*, had an eel story.

⌣

In the morning, Stella and I visited Charlie Hamlin, who lived alone in a one-room flat in the settlement of Te Hauke. Charlie was eighty-two, old enough to have lived a traditional Maori childhood, where the entire village journeyed to the river to catch eels in a *pa*, or weir, during the autumn downstream migration.*

Stella told me before we arrived that Charlie was elderly and almost blind. A bed was in one corner of his quiet room, and we sat at a table near a small kitchen with a cooking range and a sink. The numbers in Charlie's phone book were written an inch tall apiece with a black marker. There was a bucket of water next to the table.

"Mind that," Charlie said, as I had almost tripped over it.

---

* Also called a *pa tuna*, a *pa* is usually V-shaped, just like the weir Ray Turner used to catch eels in the Catskills, except instead of it being primarily built of stone, these weirs were built of wood and vines. And like Ray's weir, they were used for catching fish on a downstream migration. At the vortex of the V was a trap, usually a woven basket. Some weirs were shaped like W's with two vortices and two traps where the eels collected.

"I catch rainwater for my tea. I don't trust the water from the tap."

"Would you like us to boil some water, Charlie?" Stella asked.

"Sure, sure, that would be nice. The kettle is on the stove, there's water in it. And there's a plate of eel here my son prepared."

Stella got up to make tea. I stayed seated at the table across from Charlie.

"The frogs used to sing out here," he said, adjusting his hearing aid. "When they sprayed, the frog eggs died. They spray Roundup on the drains to keep the culverts clear so that when it rains a lot it doesn't back up with weeds and debris. But the Roundup burns the gills on the fish, and it kills the watercress, where the frogs hide, and kills the tadpoles, the primary food of the eels in our lake, Lake Poukawa. They introduced willows, perhaps for flood protection," Charlie continued, "but the willows suck all the water out of the swamps. There are not many tadpoles or native fish. The giant kokopu are gone. They're cleaning out the drains and they dredge up the mud and toss it up on the bank, and many young eels dry in the mud with it. You can't bloody explain it to the regional council people. You can't talk to the bloody *pakeha* about it, they're bloody brainless."

Charlie's vision was bad enough that he couldn't see that I was *pakeha*. But he had good reason to call the white politicians and farmers brainless.

Charlie said they sprayed the peaches with insecticide, and

the poison spread to the ground and the mowers spread the poisoned clover. He was putting posts in the orchard when he was younger and stronger, he said, and the poison caused his skin to pock up on his arms; it got in his eyes and killed all the blood vessels. That's how his eyes became white, and how he lost vision completely in his left eye.

Stella brought the tea and poured some in Charlie's mug. "How many sugars, Charlie?" she asked.

"A few," he said.

She spread a piece of smoked eel onto a cracker for him. "Not me," he said. "I eat it all the time. You guys try it." The smoked eel was soft and buttery and melted in my mouth. I had another piece and washed it down with tea. It was delicious. We sat back and listened to Charlie talk about the traditional ways that Maori fished for eels that he remembered from when he was a boy.

Where Charlie and his family went eeling, they had one of the biggest *pa tuna* around. Maori used to come from miles away and live in temporary settlements near the weir during the autumn migration. "There were other places to catch eels," Charlie said, "but they used to come here because they're the best eels you can get. Their backsides are closed up, their eyes are glazed over, they're a fat, beautiful eel. You can just about eat 'em like they are. And they're passive. Lift a big eel up like that"—he demonstrated—"he'll never say a word."

Maori centered their whole lives on the eel run at that

time of year, late January and February. They'd move their villages to the weir, "bring all their horses and pigs and all that," Charlie said. There were always a few eels that went through the weir early, before the big run, and that's what they ate while they were getting ready for the big event. In one night when the run happened, they could catch four tons of eel. "And they reckon the whole run was ten to twelve tons," Charlie said. "This is on what we called Poukawa Lake Stream. The creek was only two meters wide, but when it flooded the eels were everywhere. You can't stop 'em, you just can't stop 'em."

Charlie pushed the plate of smoked eel toward us to encourage us to eat more. I helped myself. "These are from Lake Poukawa. They're better than Whakaki eels," the ones from Walter Wilson's lake. "It's beautiful out there on Lake Poukawa. That's our playground."

Charlie explained that in the old days they didn't have cloth or fiber netting. All the traps and pots for eeling were made from a woody creeping vine called supplejack. He remembers watching his parents and relatives making their own.

"My parents, they had a net from about here to the door, made of supplejack. You can twist it around and knot it, that supplejack, it's not heavy. It grows up in the bush, and it grows fairly long. You know Maoris got a lot of spare time in those days. They sit in the sun, and they start making the hoops. They tie four supplejacks and they plait it. Them Maori, they got their way of making everything. They run out of supplies, they just go off in the bush and get some more." He laughed.

Down at the swamp at Pekapeka, Charlie said, they had two big weirs that were repaired and rebuilt for the fall migration of eels. In preparation for the run the Maori would go upstream of the *pa tuna* and clean up the whole creek, because when the flood came they didn't want sticks and things to lock up the net. When it rained and the river started to flood, they sent scouts up the creek from the weir. The scouts then signaled when the eels were coming. " 'Oh, here they come!' they'd say, and they'd get all excited." Charlie laughed and laughed with his eyes closed as if he could picture the frenzy in his mind. "You can smell eels, you know, when you pass a river and they're running." He took a big black marker and drew the site and the weir as he remembered it. "They call the baskets or eel pots *hinaki*, made all of supplejack vine," Charlie began, sketching on the newspaper in front of him, "and when the eels were coming they tied it on the mouth of the funnel-shaped leading nets, or *tawiri*, which were attached to the weir structure at the bottom of the V. They have to hold up the eels in the *pa tuna* until the full *hinaki* could be rolled away and replaced with an empty one. They have to work fast because within an hour, the whole run is gone." Charlie paused to recite a *whakatauki*, or old Maori proverb: *"Kia hiwa ra, kia hiwa ra, moe araara kit e matahi tuna,"* which roughly translates as "Be watchful, be watchful, or you'll sleep and miss the eels."

"It's a big flood, eh," Charlie continued, "and the *tawiri* fills up quick. They empty the eels into the pits, big holes on the bank, oh, ten by ten feet, and, oh, eight feet deep—there's eight of 'em down there. Each pit holds one ton. Some of the pits are still there, but most of 'em are filled in. You know, by the end everybody's down there, the whole bank's bloody loaded with Maoris from everywhere, then they have a big *hangi*, a big party. This was in the old days."*

Charlie said they stopped fishing at Poukawa Lake Stream in the traditional way, with the old weirs and basket traps made of supplejack vine, in the early 1930s, when Charlie was a boy.

"We had twelve kids in my family. My mom used to have one big plate. And she'd serve up, boil the eel up, and mix it up with potato and *puha* [sow thistle]. She used to call it *penupenu*. There was a lot of oil in eel and the oil was pretty good for you. Everyone had a drink of warm milk to wash it down. That's how she fed us in the early days, 'cause we didn't have no table and we had no plates," he added with a laugh, "just the one. And none of the eel went to waste, even the backbone. You boil the bone until it's soft, put onions in and a bit of stock, let it cool, a bit of bread and butter. That's a meal."

Charlie, eyes open, licked his lips. I imagined he was remembering his childhood home filled with the smells of his mother's cooking.

---

* The following is written in a display on eel fishing in the Auckland National Museum: "In autumn, the dawn appearance of Matariki (Pleiades) in the east not only heralds the start of the Maori new year but is also the time to get ready the *hinaki* (traps) for harvesting adult *tuna* (eels).

"Their migration downstream is foretold in the stars as Te awa o te Tuna (the river of eels) in the sky reveals itself. Te awa patahi is the start of the river. As eels move downstream they sometimes twist themselves into a tangled mass. This

⌒

In the town of Dannevirke, Stella took me to a suburban home. A large man named Robert Hape was sitting on the couch watching television. The small house was bubbling with clatter and movement, children running around the carpeted floor chasing cats and kittens. Talk turned to personal experiences that they or their acquaintances had had with *taniwha* eels.

"I had a friend," Robert began, keeping his eyes fixed on the television, "who caught an eel with a red stripe down its back. He took it home in a bucket and during the night he heard it crying like a baby. He left it in the bucket, and the next morning he went outside and saw the eel on the ground dead. Just then, at that moment, the phone rang in the house. It was a call telling him his father had died."

In the next room Robert's son said: "Sometimes we been out there, some parts in the lake, really deep, and you look and there's always one eel, and he's poking around and he'll come up to you and he'll look at you like this." He strained to bend his neck. "And I say, 'Undo the net,' and away that funny eel goes. And he knows, too. When you lift up the bag he goes straight through—see you later."

---

whiri (knot) can be seen in the cluster of stars in Te Tuna Whiri (the knot of eels). Then they move on past the eel weir te Pa-tuna to the *hinaki* (eel trap). Its mouth Te waha o te hinaki is formed by a group of stars in Te Koko (Corvus, a star), and the bottom end by Pekehawani (Spica, a star). The latter imparts energy to the eels enabling them to continue downstream whilst the ritual of mating commences, when the females select the males (Te kawao o te tairaka)."

Eels

Robert overheard us from the TV room and called to his wife, Molly, to tell us her *taniwha* story. We sat beside her at the kitchen table, the refrigerator humming. "In our *hapu*," she said, "all the men went out one day to the mountains, leaving the women and the kids behind. The *hapu* had a pet eel and the men told the women and kids, 'Feed the eel while we're away, but make sure you cut off the heads of the fish first.' Well, the boys in the tribe got smart, and they went ahead and fed the eel the fish heads. The eel got upset and left our *hapu*, and as it moved it carved a gorge through which the Manawatu River now runs.* It swam up to this place where there's a mountain peak. When the men came back they tried to beg our *taniwha* to come back, but it wouldn't. His feelings were hurt. Now whenever someone from our *hapu* goes near that peak, the peak forms a cloud over it, because the eel is weeping from humiliation. They still ask the *taniwha* if he would come back, but he won't," she said. "Hopefully one day it will."

Near Waimarama, we visited Bill Akonga, a *rangatira*, one who carries the knowledge of the *hapu*. We sat down with him in his kitchen and listened while his wife put on a kettle for tea.

Bill talked of fishing as a kid with a piece of *harakeke*, native flax—a stringy-leafed plant. He explained how they would string the piece of flax with worms, no hook, and put it in the stream. When the eels grabbed hold of the worms their

---

* This refers to Manawatu Gorge, which goes from the east side of the North Island near Norsewood, across the mountain range to the west at Palmerston North, and then to Foxton.

sandpaper-like teeth got snared by the flax and they could pull the eels in. He called it bobbing for eels.[*]

Mostly, though, Bill said, they speared eels. "There's a special place in Havelock there called Wahaparata, a little stream that flows in to the Karamu Stream. We speared there at night. The old folks used to just put their hands in the water round the bottom of the spear when they hit an eel, just so they wouldn't slide off. And sometimes the eel was bigger than their arms. They were huge things, the sort with the tusks and the dome head."

"Tell them about the barking eels," Bill's wife said. "All eels bark, don't they?"

"Well, I don't know," said Bill. "Do they?" He looked at Stella and pointed to her. "Here's the expert."

"Barking like dogs," said Bill's wife. "Just like dogs."

"I've heard they can cry like a baby, too," Stella said.

"Never heard one cry like a baby," Bill said. "But I heard them bark. Oh yeah, quite loud, oh yes. It's usually when there are a lot of them, though. I've never heard them bark when there's only one or two. Usually it's when they're fighting for food that I've heard them bark. They come out of the water—well, their heads anyway.

"When we first moved here," Bill continued, looking toward his wife, "I told her that eels bark, and she laughed at me. So we set a *hinaki* down the river, way down there. We went down

---

[*] They have a similar method for catching eels in England and other parts of Europe, with worms on a string and no hook.

there at night. Used to drive to it at that time. They were cutting trees, had a bridge put across the river. And we went down, way down to the paddock down there, and set the *hinaki*. We went back ten, eleven o'clock at night, we got there, it was pitch black. All of a sudden, we go over and I said, 'Listen for the eel.' That was the first time she'd ever heard them bark. And when we turned the lights on, whoa! There must have been hundreds of them. It looked like the water was just churning with eels. There was quite a lot of them barking. They were trying to get in the *hinaki*, but the *hinaki* was full. And they bark just like a dog. Eerie when it's, you know, pitch black. No moon, no anything."

Bill Akonga told us to go down the road to see an old friend of his. "Andrew Farmer used to smoke eels in his chimney. Hung 'em in the top—fell in when they were done. He'll spin a good yarn."

We found Farmer, a thin, frail elderly man, in his home down the road. A small tabletop radio was set beside his arm.

"In the town of Clive," he said, "they have an annual competition. Best-dressed eel. And the prize is a thousand dollars. You know, put little dresses on them, stand 'em up. John Wooky's son won one year, dressed it as a bush bug." The radio chirped indiscernible sounds.

"We didn't have a smoker," Farmer continued, "so we smoked eels in the chimney.* All the fat drips down and the fire goes up.

---

* A parallel description of smoking eels in the chimney can be found in Thomas Howard's book *The Jonnycake Papers*, about early nineteenth-century life in New England: "They [the eels] were then washed in clean sea water and hung up in the

The old Maori used to use ash to get the slime off. Maryanne's mother used to eat eel heads," he said, nodding toward his wife across the table. "Boil up an eel head, eat it all except the skull. Those big eels, the domeheads, have a lot of meat on the head. Not just the head but the cheeks." He played with the watch on his wrist.

Farmer's wife perked up.

"We were children," she said. "I was, oh, eight. And we were down the creek, and the creek went through this gully in the forest. And we heard a bark—it was loud, like a roar. Just like a dog. We went home and my mum told me it was an eel, a *taniwha*, warning us not to go down there. We never did again."

My feet were propped up on the dashboard of the car, my pen moving across the page of my notebook. Stella and I had been on the road together for a week now.

"You spend more time writing in your notebook and reading than looking out the window," Stella said. "You're in New Zealand and you'd rather live in a book. I don't think you'd experience things yourself if you didn't have to."

"That's not true," I said. "If that were the case, I would never have left home."

kitchen chimney, with its wide, open fireplace, for one night only. Next morning, the eels were cut in short pieces and placed in a gridiron, flesh side next to sweet-smelling, glowing coals, made from green oak, walnut, or maple wood."

Eels

"When you're not reading a book, you're trying to fit your life into a book."

"Oh," I said. "That's interesting."

"Ka-klunk!" Stella said, smiling.

"What's that?" I asked.

"That's the sound of a thought falling into your head."

⌐

Stella and I made our way to Wellington, where we caught a plane to Dunedin. On the way I thought about what she'd said about my compulsion to define (what is a *taniwha*?) and record what I heard and saw—to fit my life into a book at the expense of missing the beauty of the moment. There was truth in this, but I was not blind to the fact that Stella was classifying me as well; I was a Westerner, a *pakeha*.

We were flying to the South Island essentially to visit one man. Kelly Davis, in his fifties, had devoted the latter part of his life to helping protect the longfin eel. He was the Maori man who had challenged Don Jellyman at a fisheries conference, asking how tracking eels to their spawning grounds was any help to the eel.

"Nobody should ever know," he'd protested, "where these creatures reproduce."

Kelly lived in a modest one-level farmhouse with his wife, Evelyn, and two foster children, a brother and sister named Tristan and Lovey. His lawn was a graveyard of fishing equipment— nets that needed mending, an aluminum boat on a trailer, buoys and tangles of ropes. Kittens and puppies and chickens ambled around the perimeter of the house and on the driveway, which sparkled with the opalescence of crushed abalone shells.

Kelly, of medium height and heavyset, greeted us at the door in a pair of old sweatpants, flip-flops, and a torn-up T-shirt. He spoke with a heavy Kiwi drawl and wiped his longish hair away from his warm, likable face.

He invited us to sit in his living room, and sank into the depression of an off-white couch. A kitten jumped up on the couch and walked across the back of it, nestling between a pillow and Kelly's meaty shoulders. The TV was on. Kelly started right in.

"The longfin eel is my passion," he began, "my obsession. I am a longfin."

Kelly's home stream is the Waihao, but he did most of his eel conservation work on the nearby Waitaki River, one of the largest rivers on the South Island. The Waitaki has eight hydroelectric dams along its length that collectively generate 20 percent of New Zealand's electricity (75 percent of New Zealand's electricity comes from hydropower, much higher than the world average, 15 percent, or the U.S. average, 10 percent). The country and the power company, Meridian Energy, proudly advertise clean, renewable electricity generation, but the dams required to make that electricity create anything but a renewable environment for the migratory eel. "In my mind there are two main forces causing the destruction of the longfin eel," Kelly said, "hydrodams and commercial fishing."

He went on, "I remember when I first came back from overseas in the navy in seventy-three." A tattoo of a monkey on his hand and a mermaid on an anchor on his arm were relics of his days in the service. "When I returned home, our fishery was in dire straits. There wasn't the same thing you saw when I was a child. I went away in 1963, so in ten years it went from a fishery where we were walking on eels, you know, where it was hard to take a step without feeling one with your feet, to where all that sort of thing is gone."

The major change had taken place in 1958, when construction commenced on Benmore Dam, at the top of the Waitaki. It took seven years to complete what would become the largest earthen dam in New Zealand, 360 feet high and more than 2,600 feet long, creating the largest artificial lake in New Zealand and effectively cutting off any movement of eels to or from the upper river. Seven more major dams were built on the Waitaki, part of one of the largest hydroelectric power programs ever initiated in New Zealand or anywhere else. It effectively prevented any adult eels from returning to the sea and any juvenile eels from populating the river upstream of the dams.

Kelly says that when a female longfin eel is ready to migrate to the sea in autumn, she'll circle the lake trying to find a way out, and if she can't, "she'll just keep circling the lake, until the urge goes away." The eels, he says, will just keep living until they have the opportunity to get out. Or they will feel the pull of the water as it runs through the electricity-generating turbines in the dam and, taking the path of least resistance, will try to swim through the turbines and get chopped up or maimed.

The dams have turbines like giant window fans that spin horizontally as the water goes through, transforming the kinetic energy of falling water into electrical energy. The power companies don't want the eels going through the turbines, as they can damage the equipment. Meridian Energy has tried all kinds of ways to keep eels away from turbines in dams on the Waitaki—screens, high-frequency sounds, lights—to no avail. The eels feel the pull through the penstock, the tube that funnels water to the turbine. The instinct to preserve energy for their long journey turns out to be deadly.

"We've found the big migrating eels just wait for a flood and roll downriver. They don't even swim; why should they?"

Kelly hadn't thought about the consequences of the dams until he came home from his service in the navy in the early 1970s and observed what was happening. "Once the hydrodams went in," he said, "the eels had no chance of ever returning to the sea to spawn. They were landlocked."

Simultaneously, a new and lucrative commercial fishery had emerged supplying eels to overseas markets in Taiwan, Japan, and Europe.

"I was up in Twizel to interview for a job," Kelly said (the town of Twizel was originally built to house the dam workers), "and I saw guys commercially fishing on Lake Benmore there. And they were bringing up eels, without a word of a lie, that had heads on 'em like a full-grown Labrador dog. And they were cutting their heads off and throwing them over the side. And I remember going up to the fishermen and saying, 'You're going to destroy this fishery if you continue doing that!' And they said, 'Oh, they're too big, you just can't sell 'em.' So they just killed them. Why? 'Cause the big eels kept stealing the bait in their traps. And today we're wearing that fact. If they hadn't cut the heads off those eels and thrown them over the side, they'd still be in the lake today. Between dams and the commercial fishing, in thirty years the fishery had been decimated."

The big eels in Lake Benmore that Kelly observed being slaughtered were on average between twenty-five and sixty years old. Some of them, Kelly pointed out, had no doubt migrated to the upper Waitaki River years before he was even born (and many years before the dams were built).

Kelly and some Maori cousins began grassroots efforts to trap the mature eels as they accumulated above the dams during the fall migration and transport them in tank trucks to the sea. This "trap-and-transfer" program was emulated on other rivers, wherever Maori took it upon themselves to help. "It may seem futile to just move a few hundred eels in a season," Kelly said, "but you figure every big female that has a chance to get to the spawning grounds is carrying about thirty million eggs, so we know it makes a difference. That's a fish that would otherwise never get out there."*

Kelly was concerned that there weren't enough eels returning to spawn. He imagined the route the eels took to the spawn-

---

* Bill Kerrison, a Maori man we visited later in the trip, has spearheaded a similar trap-and-transfer program on the Rangitikei River (North Island)—not only to help adult eels get past three major dams on their downstream migrations to the sea, but also to help the young eels get up above those same dams. He moves more than a million young eels upstream of Matahina Dam (the largest earthen dam on the North Island) every year.

ing grounds as a trail marked by scent—that the adult eels left some trace of themselves behind that acted as signposts for the orphaned young to return to freshwater rivers.

"It's my theory," Kelly said, "that their pathway to where they migrate leaves that scent. And I've no other reason to believe it than stories that our old people talked about."[*]

A ray of sun spread through the window and across the floor onto Kelly's face. The kitten on his shoulder momentarily opened its eyes, stretched and squinted in the sunlight, then curled up and went back to sleep.

"People don't understand what value the eels are for us Maori," Kelly said. "The old people knew when the eels ran that it was time to prepare for the cold months—they were like a seasonal indicator, a calendar.

"We went seasonally to fish the longfins during what we call the *hinapouri* [darkness, or new moon], when the eels gather to migrate. We would only fish for sustenance, not to trade or anything. It was the time I liked most as a kid 'cause we had the week off school to fish eels. We fished channels in the shingle at the end of the Waihao in the Wainono Lagoon. The eels were gathered waiting for a storm so they could roll over the shingle bar.

---

[*] Eels have one of the most powerful senses of smell of any creature. According to the German ichthyologist Friedrich Tesch, author of the most comprehensive scientific book on freshwater eels, *Der Aal*: "The eel is almost as sensitive to smell as the dog, which is not surpassed by any other animal." Tesch wrote that eels can perceive the smell of roses diluted to the degree of "one ml of scent in a body of water 58 times that of Lake Constance." (Lake Constance in Europe is 39 miles long and 830 feet deep.)

"I don't know if you've ever seen 'em try to migrate," Kelly continued. "It's the most amazing sight. I've seen 'em roll up in a big ball and just roll straight over, over the beach. They go way up the river and then they swim furiously down the river, and they just roll, like a big ball, rolling over. And that's usually round about the end of April when that happens. At night you can go down there and you can see the eels, trying to work their way through the shingle.*

"The right size eels, about three or four feet, we took home [the big females were carried over the shingle bar and released in the sea]. We had a process we called *pawhara* where we open the eel along the back and clean it and hang it by pieces of flax on a wooden drying rack. Each family had a quota. Down here, for our family of six, we took two bags full, and they were huge bags—and that was enough to keep you fed for winter.

---

* Though I had never seen balls of eels, I had heard many stories about them. A commercial fisherman named Roger on the Seine near Rouen, France, told me that he once caught a ball of eels in his nets. The ball was so tight that he couldn't pull the eels apart with his hands, so he and his friend loaded the whole ball into the back of his car. When they got to Roger's house the ball of eels was still tightly woven, and he had to use a hatchet to break them apart. A 1966 paper by J. C. Medcof in the *Journal of the Fisheries Research Board of Canada*, titled "Incidental Records of Behaviour of Eels in Lake Ainslie, Nova Scotia," has several accounts of eel balls, this among them: "Frequently the migrating eels knot themselves together, and large bundles, often a fathom in circumference, are seen lying in the lakes or trundling down the streams." Bill Kerrison told Stella and me about balls of eels he'd seen as a boy in the North Island—tight, compact, and almost perfectly spherical. "I was standing on the riverbank with my grandmother, and we saw this ball of eels go by, and I asked her, 'Why do they do that?' She said, 'Because they're in love.' The Maori felt that eels formed balls as part of their courtship."

"Different people have a different affinity for eels. I mean, for us Maori, it's the highest-protein food that we can ever have in this country. A lot of things have taken over since I was a kid in terms of food, but it's still a major diet if we can get them. I take the opportunity now and again to get a feed, whereas when I was a child, we had eel at least three, four times a week, two to three times a day. I used to take the dry eel to school for lunch—it's like jerky, you know, you rip it off and eat it. It was something that sustained us, really. When money was scarce, we lived off the river and the sea; we had no other option. And we grew vegetables and stuff like that. I've been trying to tell our young people, this stuff is great for you!"*

Kelly continued to share his mountains of experiential knowledge. "The glass eels used to come in so thick they looked like an oil slick on top of the water. They come into the river year after year and hide in the stones and the watercress." When the young eels decide it's time to go upstream, after waiting in the lower river for as long as ten years, they go all at once, if necessary forming braids with their bodies to surmount vertical walls. "If one goes up the wall, they all want to go up. They're amazing." Kelly described diving in the rivers with big

---

* Indeed, a recent study published in the *New Zealand Herald* showed that Maori who consumed eel meat almost daily had fewer health problems. High in omega–3 fatty acids, eel is especially good at preventing type 2 diabetes (the form linked to obesity). The study showed that Maori who ate eel several times a week had virtually no cases of type 2 diabetes, which was otherwise epidemic among modern Maori.

eels, watching them suck the meat out of freshwater mussels. He told stories of going out on rainy nights in farm paddocks with a ferret on a leash, hunting eels traveling overland from one body of water to another. Kelly told tale after tale of the physical challenges the eel faced and of the importance of the eel for the Maori as a source of sustenance. And then he shared a more personal strain of the story: the interweaving of the eel's fate with the Maoris' own.

In the 1860s the British settlers in New Zealand established a kind of department of wildlife they called the Acclimatization Society, meant to help British immigrants acclimatize to life in the new colony.* This was accomplished through the introduction of familiar species, including red deer, pheasants, quail, ferrets, rabbits, possums, foxes, swans, ducks, and geese. From the sportsman's point of view, one of the more successful of these introductions was the brown trout, which arrived as a British import (via Tasmania) in the 1860s.

The streams and lakes of New Zealand, cold, clear, and rich with aquatic insect life, were more than suitable for trout. The speckled fish rapidly established themselves in the streams, rivers, and lakes, reproducing naturally and growing rapidly up to weights of five to ten pounds, a trophy trout for any angler. Within decades of the initial introduction, New Zealand was renowned worldwide for some of the best trout fishing in the world.

---

* The Acclimatization Society was not unique to New Zealand, but existed in other colonies of the Crown, including Australia.

But there was a strange predator in the water that British anglers were unaccustomed to seeing. Occasionally while a trout was on the line, a shadowy form would emerge from the depths and devour it. A four-to-five-pound trout was no match for a five-to-six-foot-long eel. Once the British settlers identified the giant eels as a threat to their prized gamefish, members of the Acclimatization Society set out to eradicate them. The society pinned up "wanted" posters in bait shops, offering a bounty of twopence for each tail clipping that verified an eel kill. Huge longfins were caught and left on dry riverbanks to die. Instructions to kill eels appeared on the back of all New Zealand fishing licenses, like this one from 1950:

Make War on Eels

- Excepting erosion and flood scour, eels are the greatest enemy of trout in our rivers.

- Eels are also competitive with trout for feed; when they are not eating trout they are depleting trout food.

- Therefore, every angler should KILL EELS every day he is on the water. Always carry in your fishing bag a shark hook, attached to about 3 yards of stout line. Bind the hook to the end of any stick and carry the line up the stick to your hand. Jag the eel, discard the stick, and pull the eel ashore and destroy it!

"The Acclimatization Society took as many big longfins out of the river as they could," Kelly said. "My dad and me, we used to go behind the men and kick them big eels back into the water.

This was in the mid-, late fifties. It wasn't until they found a market for them overseas that they stopped killing them willy-nilly.

"The slaughter of eels by the society," Kelly said, "was akin to what the Europeans did to the buffalo in North America. Like the Great Plains Indians relied on the buffalo, Maori relied on the eels for sustenance and for our faith. I'm not sure the Ac-climatization Society didn't know what they were doing."

Kelly's wife, Evelyn, brought a tray of tea and biscuits into the living room. Stella and I had been glued to our chairs—hours had gone by as we just listened—and Kelly still sat deep in the couch with the kitten warming the back of his head. He stopped speaking briefly to take his tea on the couch. He showed us an amulet that Evelyn had carved for him out of *pounamu*, or greenstone, in the shape of an eel. He put it on the tea tray and continued.

"There's more at stake than just losing the fish," Kelly said. "It's our way of life. We have to preserve that sense of wonder in our children, have them see one of these giant creatures. I'll stop and show children, pick a big mother eel up in my hands and let them touch it. We had some big ones in a catch basin that we caught in Lake Benmore, ready to take them down to the sea. And this car went past and stopped. I pulled this huge eel out of the bag, a big migrant female with big blue eyes, and held it in my arms. And these little kids started climbing out of the car window to have a pet at the thing. They couldn't believe it was so docile. And it was twenty-six pounds, just a huge eel, massive! And I told them the story, that the longfin eel is endemic to New Zealand, that they only live in the waters of this country, and I told them what we were doing and why we were doing it, that we were taking these big eels to the sea—it was an experience they never had."

Stella got up to get more sugar for Kelly's tea. "What's the biggest eel you've seen in your life?" she asked.

"I was up and above where the flows are upstream," Kelly said, stirring his cup of tea with a small spoon. "There're some huge eels up there. I mean, I'm talking probably twenty-, thirty-kilo longfins. I'll tell you, without a word of a lie, there's one up there, one night we were just spotlighting off the bridge—it was huge. Out of the corner of my eye was this eel coming, this thing like a powerboat, going upstream, so we shined the light on, and she just turned around, flicked her tail, and went in under the watercress. It would have been from here to that table, a good three meters long and bigger around than my thigh. And that's no rubbish. There were a number of us saw it. The amazing thing about it was that the girth continued right up through to the end of the tail, like a barracuda."

Stella asked Kelly what was the oldest eel he'd ever heard of.

"I've personally aged eels, specimens that died in the dams or the trap-and-transfer program, at over a hundred years.* The oldest eel documented was a hundred and six years, a speci-

---

* The only reliable way to age an eel is by counting the rings of a cross-section of the otolith—a small stone-like sensory organ in the inner ear—and the fish must be dead to obtain it.

men from Lake Rotoiti in the South Island. It wasn't even that big, only seven and a half kilograms. They say the eels in that lake mature at ninety-six years old!"* Kelly said, marveling. "They're just fascinating." Before Kelly had personally aged such ancient eels, he says, scientists such as Jellyman doubted customary knowledge of eels living that long.

"I mean, if they talked to us before they did a lot of things, if they talked to our people, my father and my family that lived on this river for sixteen generations . . . if they'd have talked to my father and sat down and listened to the stories he had to tell—they would have learned something.

"Our local knowledge is useful to scientists, but they just take what they want and toss the rest. I taught my girl where to get the wood pigeons in season, where the *miro* berries grow and ripen. I was out pig hunting one day and got a pig, and on the way back I stopped just after first light at a patch of miro berries I knew. The pigeons, we call *kereru*, come in and gorge themselves."

Kelly described the scene in the berry patch, where the pigeons eat so much that when they try to fly they sometimes crash-land and die. "Their crops just explode, and they can be collected on the forest floor," Kelly said. "I stuffed thirty birds in the belly of the pig and made my way out of the bush. On the way back, I ran into a guy from the Department of Conservation and he saw me with the pig and saw a pigeon sticking out

---

* Two years later I visited Don Jellyman in Christchurch on the South Island. He confirmed what Kelly said; at least, the numbers were close. Don said the mean age of the female eels leaving Lake Rotoiti to spawn is ninety-three years.

of the pig's mouth. He wanted to pinch me because the pigeons are protected. I said, 'I'm allowed to do this, it's my customary right,' and I said, 'Besides, I didn't kill them; I just picked them up off the ground. They killed themselves.' He didn't believe me. He said, 'We need to know where this place is,' so I took the guy and showed him the pigeons, dead all over the ground. He was amazed. I said to him, 'That's why they taste so lovely this time of year, they're so fat with berries.' "

And still, not knowing certain things had always been of equal importance to the Maori as knowing—for instance, where the longfin eel goes to spawn.

"Look, mate," he said, "I prefer that they leave them alone. That's my feeling. I mean, they're very interested in finding out where this house is where they breed. Why? To benefit the commercial interests?' "

Because New Zealand eels are some of the largest in the world, they can accommodate a sizeable tracking device, and therefore offer the best opportunity to track eels to the spawning grounds given the existing technology. Don Jellyman placed a tag onto each of ten large female eels (using a nylon bridle to hold them into place) and released them into the sea near Lake Ellesmere on the Canterbury coast of the South Island.*

---

* The tags could be timed to release from the eel and, being buoyant, come to the surface and transmit data of the fish's whereabouts and daily activities via satellite to a computer. Don's first effort to tag eels, in 2001, did not yield a great deal of information about where the longfins went to spawn. Perhaps the most intriguing bit of data retrieved from one of the eels was its pattern of traveling in an undulating formation, swimming near the surface at night and diving to nearly three

"Jellyman's work didn't come to anything," Kelly said, pleased. "After all that money"—each tag cost $4,500—"the tags and the eels went missing. That money could have been used for trap-and-transfer, to help the eels around dams. How much more do you want to study them? The research that they've done hasn't been beneficial to the eel, that's my view."

After hours of talking, and some tea and biscuits put away, Kelly drove us to his local *marae*, the Maori tribal meeting place.

"When you come back," Kelly said, "you stay here. You just come and be my shadow and I'll show you things you've never seen." He was standing on the steps of the *marae* in a heavy wet drizzle. We walked around the back of the *marae* and he showed us a *waka*, or sea canoe, he built with the local children, made entirely out of bundled reed grass. When I asked if it was seaworthy, he said, "You bet." He told Stella and me the story of how the South Island was actually the *waka* that the first New Zealanders had taken to this area. The canoe beached on the South Pole, they say, on the shoals of Antarctica. Then Maui, the leader of the expedition, fishing from the beached *waka*, pulled up a giant stingray, and that became the North Island. And if you look on a map of New Zealand, the South Island

---

thousand feet during the day. Don later told me that the reason for this behavior could be to evade predators or delay sexual maturation in the cold depths until the eels get to the spawning place. Similar travel patterns were observed in results from a 2006–07 study tracking European eels from the west coast of Ireland to the Sargasso Sea.

does resemble a canoe, and the North Island does look like a stingray. Kelly told the story not as if it were myth, but as it was to him—the beginning of his genealogy as a Maori. The rain fell harder and we got back in the car.

chapter four

More Tales
OF *Taniwha*

Stella's father, Falla August, with a longfin eel

I t is hard to see freshwater eels in their habitat. Oftentimes they live in murky water, are active mostly at night, and flee if approached. It is an unusual character trait of the New Zealand eels that they come out willingly to feed during the day and can be found in rivers and streams that are crystal clear. This offers a good opportunity to observe them in their element. I had brought some snorkeling gear and told Stella I was determined during our trip to watch eels underwater.

After visiting Kelly we returned to the North Island. Stella said she knew a good place to see big eels on our way back north to Napier and her family farm, and that I could try out my snorkeling gear there.

Standing on a bridge, peering into the creek at Mount Bruce Wildlife Preserve, I watched Stella toss mackerel chunks into the water to coax the eels out from the undercut banks. As the huge eels came out from the dark shadows, I was stunned by their enormousness, their sleek mysterious movements.

"Having second thoughts?" Stella asked.

I walked carefully down the steep muddy bank to the creek and eased into the frigid currents wearing a mask, snorkel, and wet suit. As I swam among the big eels, they came up to me and bumped their noses against my mask. It was hard not to flinch as one approached, torquing its muscular body, thick as the calf of my leg and close to five feet long.

How to define it? Was it an ocean fish or a river fish? At that size, bigger than an average large salmon, it felt to me more like an ocean fish. Yet how much history it had with the land—the seasons, the storms, the farm animals, the stones and leaves, the bird songs, the insects, the daily cycle of the sun, things tangible to us. In the ocean it was capable of traveling in the aphotic zone—at depths where no light penetrates. How did it navigate? The eels' motion in the pool at Mount Bruce Creek was beautiful, lyrical, sinuous, symmetrical, but also fearsome. It was a sacred moment being in the water, body fully immersed with them.

The images of the large eels lingered in my mind as we

headed to the farm where Stella had grown up, on an isolated stretch of coastline near Hawke's Bay. Over the next few days she and her family would be throwing a big party to celebrate the sixtieth birthday of their good friend Papa Bear.

⟶

The land at the end of the dirt road was broken and hilly. Through a haze of sea mist there was ocean, and between the road and the waves was a clutch of buildings half hidden by bushes and trees whose tops had been sculpted into teardrop shapes by exposure to the persistent southerly winds.

Stella's mother, Shirley Cunningham, an Englishwoman with a face channeled by the wind and sun, greeted us, smoking a cigarette. With little delay Shirley enlisted Stella to help corral sheep into a pen before dinner. Stella jumped right into the mud and helped her mother push and pull the wet animals until the job was done.

Stella's family had moved to this piece of land from town when she was seven years old. As a girl she took to the untamed surroundings, developing an affinity for the sea and rivers. She and her father gathered *kaimoana*, seafood—diving for *paua* (abalone), pulling up crayfish pots, and collecting *karengo* (seaweed) and *bubu* (whelks) from the rocks at low tide. She learned to always respect Tangaroa, the god of the sea.

When she was eleven, Stella's mother left, and the girls were alone on the desolate coast with their father. In those years her knowledge of the natural world only increased, as did her sense of independence. Two weeks before Stella's sixteenth birthday, her father was killed when a rogue wave overturned his boat while he was launching it from their beach. Stella's last memory of her father was helping him into his wetsuit the morning he died. Devastated, the two teenage girls asked their mother to come home and look after them. Stella's mother had lived at Kairakau ever since, tending to the sheep and cows with her boyfriend, Ray.

⁓

Ray was excited to hear about my interest in eels and suggested we set a *hinaki* in the creek that night. At twilight we went off in his truck along a treacherous and deeply pitted road to a wooden shack. Ray said an old Scot used to live there, who made his living culling rabbits for the government. Leafless pink flowers with single stems jutted out from the ground like lollipops.

"'Em is called naked ladies," Ray said, "'cause they have no leaves." Ray pulled the wire eel trap out of the bed of the truck, put in two fresh sheep livers, and secured the top.

"The old Scot fished this corner 'ole in the creek. It used to be good eeling 'cause they dumped dead livestock there."

Ray lobbed the trap into the hole and we watched it disappear as it sank to the bottom. He tied the trap to a tree and we headed back to the farm.

After a dinner of lamb cooked on the charcoal grill, I pitched my tent near the beach and got in my sleeping bag. I listened to the surf pound from my cozy place, the wind whooshing in the trees. I rolled up my jacket to use as a pillow and fell asleep.

Early the next morning, Ray and I returned to the rabbit

hunter's shack to pull the eel pot from the creek, but all that was in it were two wet sheep livers. By the time we got back to the farm the guests had started showing. Throughout the day family and friends arrived in droves, pitching tents, parking their trailers, cooking food, playing music, transforming the desolate beach into a party ground.

A friend of Ray's named Carl saw our empty trap in the back of the truck and said he'd help me catch a few eels in the creek that night.

Carl and I started drinking beer in the early afternoon along with everyone else. By dusk we were all pretty buzzed. Carl suggested we set the net before we drank too much and before it was dark, so we hiked up the beach to the mouth of the creek and then further upstream to scope out a good spot.

Carl's trap was different from Ray's, made of netting attached to concentric wire hoops—a fyke net. He also brought different bait: the tail, head, spine, and entrails of a sea fish called a gurnard.

The creek was very winding and off-colored, which apparently is how it always was. Carl rolled a cigarette and puffed while he read the currents.

"See," he explained, "this spot's no good 'cause it's a mud bank. But up there the bank is hard, and the stream goes under the bank. Now, *there's* a spot an eel would like. You got to think like 'em, eh? If you were an eel living there and there was a flood, you'd have a chance of not being washed down. If I gave you a choice, would you live in a shelter made of paper or cement? It's just common sense."

I watched Carl as he began to lay out the net in the creek. "I set the net in the fast water," he said, "so it's harder for the eels to suck the bait through the mesh. They feed by sucking, and when they grab hold of something they spin."

Carl set the net with the mouth facing downstream, in the riffles at the top of the pool. The net was cone-shaped when extended. He staked the net to the ground on the bank, he explained, so a big eel could not pull it loose.

"We only just started fishing this creek," he said, putting the last stake in the ground. "It's loaded, mate."

On the way back in the dark, we walked along the stream bank on sheep trails through *tikouka,* or cabbage trees. I was startled at turns by magpie calls and the wing-clapping of wood pigeons roosting for the night. We made our way out of the trees and toward a berm that separated the windswept beach from the wood.

As we walked I asked Carl if he'd ever seen a *taniwha.*

"I don't believe in all that superstition," he said. "Many of the people who eel a lot, Maori and *pakeha,* don't go for the legends." Yet a few steps further along the trail he began to recount a strange eel experience from his boyhood.

He was fishing a tributary of the nearby Tukituki, the river where Stella had done the research for her master's thesis. It was spring and the glass eels were running up the river in force.

"I was just a little weaner," he said, "and I was scooping glass eels with a jar from the creek, just dip the mouth of the jar and suck 'em in. And it's weird—I turned around, there was no reason I should have except I had this feeling, and there was this clear eel holding in the current with its tail cocked. It was clear as glass, and fat and about as long as my leg. All you could see was some red below its head, like his heart. The tiny glass eels were coming up and this thing was coming downstream, like it was guarding them. I felt like the eel was trying to warn me not to take any of the baby eels coming upstream. I didn't."

Back at the party grounds, the group had erected large tents and rigged lights. Carl and I were hungry, since we'd been away two hours, and we were happy to see, and smell, a huge banquet laid out. It was a real *hangi*, as Stella called it—*paua* fritters (the abalone collected right off the beach that day), crayfish (like a lobster, but no prominent claws), steamed green mussels, pork, lamb. Everything came from the sea or the paddock and was prepared specially for the party.

Papa Bear's friends got up and toasted him. A few people addressed a recent political speech by Don Brash, leader of the National Party, about ending preferential treatment for Maori. The most pressing issues were related to traditional fishing rights to the foreshore and seabed. Those of Maori heritage claimed that under the Treaty of Waitangi, that right had never been passed to the Crown. Brash's speech had created new waves of resentment for Maori—an attempt to polarize New Zealand-ers in hopes of winning the next election against the incumbent Labour Party.

"But here we are," one of Papa Bear's friends said, "Maori and *pakeha* together, in this beautiful place."

Sometime in the dark and damp morning hours, I woke to find that I'd fallen asleep on the grass under the big tent. I stumbled to my tent and crawled into my sleeping bag. By the time I woke again it was light, and my tent was warm from the sun. Carl and Ray had already pulled the *hinaki*, which overnight had filled up with eight big eels. Ray had the eels alive in a large plastic barrel and was about to pour powdered laundry detergent on them to kill and deslime them.

Eels can be killed and prepared many different ways. Back home in Connecticut fishermen usually remove the skin by cutting a ring around the head and pulling it off like a glove off your hand. People who smoke eel usually leave the skin on the fish but remove the unappetizing slime with salt, ash, or detergent.

Even Ray, who had worked in slaughterhouses his whole life, admitted their method was cruel—removing their protective slime while suffocating them.* The eels reacted by writhing and rolling in the dry detergent, trying to use their tails to get it off their bodies, but it only spread the caustic powder.

When the eels were dead Ray rubbed them down with a sugar sack to remove any leftover slime, snipped their tails to bleed them, and hung them by their heads. After they had hung for several hours, Shirley took the task of splaying open the eels for drying.

---

* According to the German ichthyologist Friedrich Tesch, eels "can meet virtually all their oxygen needs cutaneously," meaning that if they must, they can survive breathing exclusively through their skin. They can only do so, however, if the skin is wet—covering them with dry powder will certainly kill them.

She made an incision along each side of the long dorsal fin, running the knife blade along both sides of the spine until the flesh was free of the bone. The guts came out with the head from the back, the belly skin holding the two fillets together. The *pawhara*'d eel was then heavily peppered and hung to dry.

Since the assorted members of the camp didn't have any manuka, the preferred wood for smoking eel, Papa Bear offered to fry some. He sliced the eel into chunks and cooked it in a skillet with beef tallow, skin side down. He said he usually ate the skin, but only if he'd cleaned the eel himself. He did not trust other people cleaning it. "I don't approve of the use of powdered detergent to get the slime off," he said. "I prefer wood ashes."

A few people had gathered around the skillet, watching expectantly as the eel steaks turned golden brown in the fat.

"The skin fries up like pork cracklings," one elderly Maori woman said, smacking her lips.

I asked Papa Bear if eel was served in New Zealand restaurants.

"Ah, nah," he said, "it's just not done. I'm not sure why they're thought of here as second-class fish, maybe because the Maori eat them." He laughed. "More for us."

In the afternoon sun the red-faced *pakeha* and the deep-brown-skinned Maori began to prepare for yet another night of partying. They cleaned out the grills, washed pots, picked up beer bottles, brushed their teeth, and shaved in small mirrors leaning on shelves over washbasins. I half remembered, as my hangover lifted, playing guitar and singing the night before. The green grass was trampled like playing fields after a tailgate, and damp under the tents, and the sea breeze was suffused with a crushed green smell. The eels we had not yet eaten still hung on the line in the sun.

A Maori bush guide who resembles a cross between a cowboy and an Indian, Daniel Joe is tall and lean, with a long nose and a casual swagger. He wears a silver pocket watch in a leather pouch (wristwatches, he says, get hung up in the bush) and a pig-hunting knife in its sheath side by side on his belt.

To meet up with DJ, Stella and I drove from Hawke's Bay to the region south of Lake Taupo where he lives. I'd first heard of DJ and his home, called Double Crossing, from my friend David Seidler—they had fly-fished together for years—and it was through DJ that I had come to know Stella in the first place. By this time in our trip, I'd heard enough stories about him—the loner, the adrenaline junkie—that he had taken on a kind of mythical aura, as had his cryptically named home on the Taupo-Napier road.

I'd imagined that DJ's place was called Double Crossing because some bad deal had gone down there—that someone had died in a shootout, Old West style. The name had arisen, though, from the simple fact that in order to get to DJ's home you had to cross a pair of rivers. The first crossing is a small stream that can be driven through with an all-wheel-drive car at most times of year. The second is a formidable river, the Waipunga, and there's no bridge, only a cable and a small plywood car.

When he first gave up years of working in the freezerworks

(the colloquial term for a slaughterhouse), DJ moved to Double Crossing, a piece of tribal land he inherited from his grandfather. No one had lived there since the early 1900s. There were no buildings or amenities on the site and no apparatus for crossing the Waipunga. For the first year he lived in a bus that he towed across the river with his friend's tractor. Whenever DJ came or went, he had to ford the river on foot (waist deep if the water was low) or take his horse across. When any of his bros came to visit, they brought DJ something—a chair, a mirror, a window, a mattress. He lived on pigs from the bush behind his house that he hunted with his mastiff dogs and a knife, trout and eels from the river, and squash, lettuce, and *kumara* (sweet potatoes) from his garden. Eventually he brought out a small prefab building, arranged some furniture, built bunk beds, and all but concealed the building by transplanting full-grown *ponga*, or tree ferns, around it. DJ started bringing timber across the river, one plank at a time, to build a larger home of his own design, while the bunkhouse slept occasional visitors. He put cement footings on either side of the river to support a cable across it, and outfitted the cable with a cage on a pulley so he could traverse the river with some supplies, or his dogs, without getting wet—such a cable car in New Zealand is known as a flying fox. Double Crossing had been revived.

When he's not guiding fly fishermen, white-water rafters, or Korean and German pig hunters ("the Koreans are in it for the gallbladder," DJ said), there are times of year when DJ won't leave the Crossing for weeks. He has a radio that runs on batteries (when he has batteries), from which he gets BBC news tailored for New Zealand, but there is no electricity at the Crossing, no TV, no telephone. His girlfriend, Nikki (Stella's cousin), was visiting with her kids one day and the kids were talking. "What's that you're on about, plane bombs?" DJ asked. The September 11, 2001, attack on New York had happened two weeks before.

Stella and I pulled off the road as directed, at the Rangitaiki Pub. DJ was having a beer and "a yarn with the publican." He invited us to join him, asking for three more Tui beers. He hugged Stella and we told him a bit about our trip so far.

"All right then," he said, "let's go."

DJ led us to his driveway a few miles down the road. The entrance was well hidden—a treacherous left turn. We followed the pitted road through the first river, water over the bottom of the doors of our rental car, until the track dead-ended on the banks of the second river.

DJ helped us unload our gear from the car and whisked us across the rushing currents of the Waipunga in the flying fox.

"I call it Air New Zealand," he shouted as we flew.

Double Crossing was a farm of sorts, skirted by native bush and embraced by a horseshoe bend of the river. DJ's home was ringed by large tree ferns, and behind was a grove of kahikatea trees over a hundred feet tall. In the uppermost branches large wood pigeons cooed. Several horses grazed on the flat green common, and his three bullmastiff-greyhound dogs greeted us with tails wagging.

"Hey there, Football-head," DJ called to one. "Hey, Nunu, hey, Bruiser."

DJ helped Stella and me settle in the guesthouse, concealed by tree ferns and vines. Inside there were a few bunk beds, a kitchen, bookshelves, a couch, and a dining area. Wood figures adorned the shelves and available counter space, one of a twisting eel DJ had carved, its mouth and eyes natural features of the burled wood. A resident cat named Pussy Galore was asleep on the couch, curled up in a slanting ray of late afternoon sun. And on a windowsill above the couch were a pair of pig skulls.

DJ popped a beer for each of us, and we sat on the porch looking off across the farm.

"I don't take many trout from this river," he said. "They taste a little like clay, but they're okay smoked over manuka wood. What do you say we catch a few for dinner, eh? Then we can use the heads to bait the *hinaki*, leave it in the river tonight, see if we can catch a few eels."

Stella stayed behind at the house while DJ and I hiked upstream with our fly rods. We drifted our flies through the cold, milky currents of the river and in an hour of fishing managed to get three good-sized rainbow trout. I strung each of the silvery fish through its gills on a supple willow branch, and we carried them back to Double Crossing. On the way we stopped at a blackberry patch and got our fill of berries. DJ pointed out the call of a tui bird in the bush and said he often heard the mechanical song of bellbirds in the mountains behind the house.

Back at Double Crossing, we hung our waders on the porch of the guesthouse and put away our fly-fishing gear. I filleted the trout while DJ lit a fire in the fire pit to smoke the trout over manuka chips. Stella came outside the guesthouse to help prepare dinner.

"Maori traditionally smoke fish with manuka," she said, watching DJ prepare a metal plate with shavings of the wood. "Honey from manuka flowers is renowned throughout the country."

"There are two methods commonly employed for smoking fish," DJ instructed, "hot smoking and cold smoking. Cold smoking is done when you have time. Hot smoking is done when you're hungry.* Maori used to cold-smoke the heads of their enemies and relatives."

Stella began preparing salad and boiling potatoes on the gas stove inside the guesthouse. While the trout was cooking DJ poured me a glass of wine and then put me to work mending holes in the eel pot we would set that night. DJ said that if every hole wasn't plugged, the eels would find their way out, tail first. My mending passed inspection. In the meantime, the trout was cooking on the grill over the smoldering manuka chips.

DJ suggested we set the eel trap before dinner, otherwise we'd have a hard time walking back along the river in the dark. The heads from the trout we'd caught went into a mutton-cloth sack that was tied to the far end of the trap. The eels could chew on the bait inside the mutton cloth but not get at the bait itself.

---

* Cold smoking is done at a lower temperature and relies on the smoke to cure the meat, as opposed to hot smoking, which in effect cooks the meat. Cold smoking preserves meat for longer periods of time but can take hours or days.

When the trap was wired shut, DJ and I carried it upstream from Double Crossing, threw the trap into the river, and tied it to a tree on the bank.

It was now dark. A rising moon had lit up the riverbed and brought shining highlights to the gurgling riffles of the river. When we returned to the Crossing we opened another bottle of wine and DJ added a few more logs to the fire. We watched the fire in silence.

DJ poked the fire with a stick. The flames illuminated the strong angles of his face. "The way I see it, there are three players in the river: the rainbow trout, the brown trout, and the eel," he began. "The trout are British imports. The eel is the cultural factor. Everyone forgets about the eel because you don't see them. You don't see the eel, but he's there, and he's *relentless* in his efforts to catch the trout. He's always stalking them. Ultimately, *he's* the survivor. He can take the other two out anytime. He might wait years to catch them, till they get old and weak. The eel, old Tuna, he's got time. He's been there before the trout and he'll be there after. We call that *morehu*—the survivor.

"I liken that to our present situation in New Zealand," DJ continued, "between the Maori people and the Crown. The Maori people now are the *morehu*. We're in pursuit of correcting the wrong that the Crown brought to us. Stella and the ones who are getting educated in the universities, they are told by their elders, like I was told by mine, that we have to fight. Our land is our economy, and the last foothold is the foreshore and the seabed. They want to make it legal through legislation, to take those rights from us." DJ paused to drink some wine.

"Eh there, Stella?" he said to her. Her long black hair looked especially dark this night.

Stella responded with a sedate "Yeah."

I looked at the largest of the trout, its flesh a beautiful orange-gold color. Back home this trout would have been a trophy; here trout this large are commonplace. Ironically, while the large size of New Zealand trout could be attributed to a rich food supply and mild winters, another contributing factor was the eel.

The Acclimatization Society was correct to think that the eel was a predator of the trout—eels eat plenty of them. But in the years that the eradication efforts continued and anglers and society members killed thousands of large eels, they saw their trophy trout fishery change. The trout in eel-free rivers had become more numerous, yes, but the average size was much smaller.

In the 1950s, a biologist named Max Burnett studying the interaction of trout and eels in the streams of Canterbury discovered that the eel, maligned and needlessly slaughtered, was actually in part responsible for the now world-famous trout fishing in New Zealand. By preying on the trout, the eel was culling a population that soon became overpopulated and stunted without them. With the eels in the rivers, the trout were fewer but much larger. Burnett's work showed that the presence of eels was beneficial and single-handedly turned around public opinion of them. The killing stopped.

A young technician working with Burnett on the study, who as a child had killed eels for money, became fascinated with the life history of the snake-like fish, and in the process he has become one of its greatest advocates, not to mention one of the world's best-known eel scientists. His name is Don Jellyman.

The fire burned to embers, the night grew cold, and we moved inside to eat dinner at a long table lit by candles. The tabletop itself was a conversation piece, a single slab of wood about twelve feet long. "The tree was from the property," DJ said. "It's an endemic called matai. My friend Alex, a helicopter pilot, had to fly in an old portable Patterson saw to mill it. We cut a hole in the wall with a chain saw to get the tabletop in the building."

We picked at the meat on the trout skins. The flavor conferred by the manuka was light and sweetly smoky. Football-head, for all his intimidating massiveness, was tranquil beside us, chewing on his cow femur. DJ poured us another glass of wine. Stella asked DJ if he'd ever had a run-in with a *taniwha*. DJ picked his teeth with the end of a flat matchstick.

"My friend Alex and me were deep in the bush hunting pigs," DJ said. "We went up on horseback, so we had our horses, and we were pig hunting, so we had our dogs. And we'd camped under this permanent shelter, a hut. It'd been there forever. We'd had a long day of hunting and we were cooking a big feed. It was getting toward dark, like now, and all of a sudden the bush went silent." DJ paused. "Normally the bush is full of sounds at dusk. Well, it all went dead quiet. Then the horses started acting up and the dogs went ballistic. We're told as kids, 'Don't ever camp on top of the track,' but we did—we were set up right on it. I'm always trying to reason, and I thought, there's some logical reason why they're acting up. I thought it was possible that an experienced horseman was riding up in the dark, up the track along the steep ridge that we'd taken up there. People do it all the time, it's possible. I waited for that horseman, but he never came."

Stella looked into the candle flame on the table.

"Well," DJ said, "now we were up on our feet and we went for cover. And then this roar came, a deafening sound. I don't know how else to describe it, and I never heard it since."

I momentarily thought of the eel trap sitting on the bottom of the frigid, dark river, and I was glad that we were warm and safe in the guesthouse. Before I fell asleep I wondered if any eels had found their way into the trap; if they had, I hoped they weren't *taniwha*. At some point DJ left the room and went back to his house ringed by *ponga* to sleep.

At five-thirty in the morning I heard DJ rattling pots in the kitchen of the guesthouse, boiling water for coffee. He was waking me up because we had to pull the *hinaki* at first light. DJ maintained that if you pulled the eel pot after the sun was up, the eels would escape the same way they came in.

It was a cold morning and dark—summer was winding down (by this time it was late March). I could still see the odd star in the sky. I put on my waders and followed DJ down the horse track to the riverbank, through the blackberry brambles where we'd gorged ourselves on fruit the night before. Searching in the early platinum light, DJ found the rope that attached the *hinaki* to a tree onshore, untied it, and hoisted the trap out of the river. It had four eels in it, and one of them was big.

"It's a good haul," DJ said. "It's all of five kilos that one, eh. It's heavy. Why don't we each grab one end of the *hinaki*." In my eyes the eel was a monster. I looked at it hard to make sure it didn't have any red eyes or stripes. It was all dark brown, dark eyes, a fish of the night, mysterious and moving its muscular body forward and backward with equal facility.

Once we were back at the horse corral, DJ began the process of cleaning the eels by putting them in an old sugar bag with ashes from the fire pit. While the ashes were desliming the eels, DJ suggested we have some breakfast.

Stella was out of bed and joined us for lamb chops and a can of spaghetti with toast. Bruiser and Nunu lay half asleep at our feet.

Stella, still in her pajamas, cleaned up some of our mess from the night before. DJ and I went out under a grove of giant macrocarpa trees and overturned the sugar bags, spilling the ash-covered eels out in the grass. The biggest one started moving through the grass, white ash glued to its skin. DJ took his pig dagger out of its sheath and handed it to me.

"You do it, James," he said.

"Stab it through the skull?" I asked.

DJ nodded.

I did as I'd seen eel fishermen do in Europe—just stick them in the top of the head with the point of the knife. I stuck the three smaller ones, but when I got to the big one I turned away.

"I can't, DJ. You do it."

"No, James, I think you ought to," he said, and looked at me coldly. I felt a shiver of vulnerability even though I was the one holding the knife. The big eel made its way across the grass, covered in leaves and dust and ash, downhill toward the river.

"I can't, DJ," I said again, turning the knife and extending the handle to him. He wouldn't take it.

"No, James, you've got to do it all yourself, mate."

"But why won't you do it?"

DJ laughed his hard heckle. I felt like I was still asleep and dreaming.

"Why?" I asked DJ again. "Have you ever killed such a big eel?"

"I've killed them bigger than that."

"Then why won't you do it?"

He looked at me again and smiled his sparsely toothed smile. It wasn't a friendly smile.

"I reckon it's all yours, mate."

I looked at DJ again. He looked at the ground.

I walked across the grass and took the big eel in my left hand. My thumb and forefinger went only halfway around its girth. I grabbed it right behind the pectoral fins, pinned it to the ground as it did its best to squirm away, took the pig sticker in my right hand, and stabbed the eel between the domes of its fat head. I handed DJ the dagger and he wiped the blood off the blade on the grass.

The eel continued to crawl across the ground.

"How many pigs have you killed with that?" I asked DJ about his knife.

He didn't answer right away. We were both focused on the eel.

"I reckon you should stick it again," he said, "but further up the head."

So I took the knife and plunged it again, but it met with resistance. I pushed the knife deeper through the skull, heard it crack through, and felt the blade scraping in the sandy soil beneath.

"The rest is just nerves," DJ said.

Stella came down from the guesthouse and we walked the big eel over to DJ's house, where he had a scale to weigh it. It weighed about fifteen pounds.

"How old you reckon that eel is, Stella?" DJ asked.

"About sixty years," she said.

It made me feel bad. Knowing it must have been old was one reason I hadn't wanted to kill it—that, and my intense and unmentionable superstition. DJ patted me on the back.

"It's okay," he said. "It's food, James. I'm going to give it to the old people, who are too old to set their own *hinaki.*"

I looked at the eel hanging there, and with a piece of newspaper pulled the slime and ash from it. I had never killed anything that was older than I was. But, having done it, I felt unofficially part of the Maori awareness. And I swear, as I looked at the eel I felt a certain clarity—my vision became clearer. I don't know how else to describe it except that I felt enlightened.

Yet, simultaneously, my body flushed with regret. I couldn't help but feel that somehow DJ had tricked me into acting out the parable of the British treatment of the Maori culture, even as he'd asked me to do so in order to be inducted into his culture. It was complicated and weird, but in the end it didn't all need to be explained.

DJ cut the eel's tail to bleed it out. Blood dripped into the pit he'd dug around his house, a kind of moat. DJ said he intended to divert water from a nearby spring to fill it up, so that it would be like a pool in a creek you might see in the bush, with ferns overhanging it. He'd gotten the idea when he was building the house, he said—a vision. As he dug the pit, he thought of the trout he'd put in the pool, and then he'd drop in a big eel "that would stalk and eventually kill the trout."

"When the Europeans came, they introduced the trout," said DJ. "Then what happened is, the trout ate all the small native fish, the kokopu and the bullies. They thought to themselves: 'Right, we own it now, we're kings of the pool.' And then, from out of the depths, from the darkness, came the cultural factor, the old *tuna*—the giant eel. He's an old fish, and he's absolutely relentless, and he relentlessly stalks the trout." DJ paused. "The eel is *morehu*, the survivor. I think they'll be there till dot. Till the end of the world as we know it."

At the end of my trip to New Zealand with Stella, I felt as though I'd been immersed in a place where not knowing is something quite different from ignorance—where the unknown is tangible and sacred, whether it be the force of a water guardian or the spawning place of the longfin eel. The nameless, the unclassified, the interstices between visible objects—all these existed in their own dimension some distance apart from concrete evidence (which by the nature of its conclusiveness seemed false and simplistic). If there was anyplace left in the world where mystery was palpable, where unknowingness was implicit, it was New Zealand. From the day I left I looked forward to going back.

chapter five

THE FIRST TASTE OF

Freshwater

In late winter, about mid-March, when plates of ice still line the muddy banks of tidal creeks, baby eels are just making their way from the sea into freshwater streams of southern New England.* From Little Compton, Rhode Island, to Manhattan Island, tiny fish, small as matchsticks and clear as glass, ascend the estuaries and lagoons invisibly, moving from salt to fresh.

A year or so before, between February and April, these fish were hatched from eggs in the Sargasso Sea. They drifted and fed and grew in the rich seawater, sheltered in their early days in thick mats of sargassum weed and by their transparency.**

---

* At this time of year in New Zealand and other parts of the Southern Hemisphere, their autumn, adult eels are beginning to migrate to the sea to spawn.

** How long it takes for an eel larva to get to the coast of North America from the time it hatches is still speculative—no one has been able to follow a juvenile eel on its ocean journey to know for certain. It is thought that it takes a larva two to three years to reach the coast of Europe from the Sargasso.

Riding currents flowing to the west, the first eels entered estuaries and freshwater rivers of the Caribbean Islands about November. By January larvae were making their way into the Gulf of Mexico and toward the east coast of Florida. Subsequently, they reached estuaries and bays of the Carolinas and Virginia and Maryland. By March, April, and May the rivers of southern New England, Maine, and Canada were receiving their manna, including the stream that connects the sea to the pond across the street from my home in Connecticut.

The ocean spawning location is strategic—at the head of a particular current, a river in the sea, which ensures that the hatchlings will be carried back toward the coast. How the young eels know to move west out of the ocean river to enter the freshwater rivers of North America is a mystery—if they didn't, they would continue riding the North Atlantic Drift, like corks in a river. And how some eels enter a given river and others move on to populate different rivers, distributing themselves more or less evenly throughout the range, is also a mystery.

By the time the eels first enter freshwater from the sea they have slimmed from their leaf-shaped larval stage, though they are still totally transparent except for two black dots for eyes. The glass eels move up the estuaries at night on the incoming tide, largely unnoticed by humans as they swim past, under, and around our businesses, homes, dams, and bridges. Days after entering freshwater, the clear fish become pigmented, like thin black shoelaces, at which point they are called elvers.*

The juveniles feed and grow and become resident eels in an estuary, river, or lake, occupying a particular place for ten to one hundred years, depending on the species. At this stage they are sometimes referred to as yellow eels because they are usually yellow-brown to olivaceous in color. When they are of age to migrate and spawn they undergo physiological changes that prepare them for their long sea journeys—their eyes get bigger and take on a bluish cast, their skin becomes thicker, and their pectoral fins elongate. Steel-colored, with black backs like many ocean fishes, these migrating eels, fat and strong, are referred to by biologists and fishermen alike as silver eels.**

---

* The word *elver* is thought to have come from a phenomenon in mid-May on the Thames that eel fishermen used to term the "eel fair." At this time of year many accounts describe (and this was not unique to the Thames) a run of young eels, each about two inches in length, that formed a dense column five inches wide that ran uninterrupted for many miles. The following account is given in *The Natural History of the Fishes of Massachusetts* by Jerome V. C. Smith, 1833: "As the procession generally lasts two or three days, and as they appear to move at the rate of nearly two miles and a half an hour, some idea may be formed of their enormous number." Smith continues with a telling account of how the eels divide their numbers as they move past tributaries of the main river: "When the column arrives at the entrance of a tributary stream . . . a certain portion of the column will continue to progress up the tributary stream, and the main phalanx will either cross the river to the opposite bank, or will, after a stiff struggle to oppose the force of the tributary branch in its emptying process, cross the mouth of this estuary, and regain its original line of march on the same side of the river." Unfortunately, such epic migrations no longer occur on the Thames.

** The distinctiveness of the eel's appearance in its various life stages led to the creation of numerous species names for the same fish. In Europe alone there were

Eels

Eels once traveled up the Mississippi and tributaries as far as Iowa, Ohio, Minnesota, and Illinois, in numbers significant enough to support commercial fisheries. In New England old-timers talk of "slicks" or "rafts" of glass eels moving up tidal creeks in spring, so thick they formed mats on top of the water. To overcome obstacles such as vertical walls or waterfalls, they would form braids with their bodies, a phenomenon known by old Mainers as "roping up." These days an eel caught in the Midwest is an aberration, and many say that such huge runs of glass eels don't happen anymore. The range of the American eel is shrinking, the total population declining, and while some attribute it to habitat loss and pollution, others blame a commercial export fishery for glass eels to Asia that developed in the late 1970s but did not hit its peak until the mid-1990s—an event that fishermen and conservation officers alike refer to as the eel gold rush.

The international trade in eels, a multibillion-dollar industry, is driven largely by Japan's appetite for the rich, fatty flesh.* The

---

over thirty species names given for what is now considered a single species of eel, *Anguilla anguilla*.

* According to Tesch's work on eels, total world eel harvest from fishing and aquaculture in 1995 amounted to over 205,000 metric tons, with an estimated market value of $3.1 billion. Eels represented 12 percent of the total value of world aquaculture production that year. A single eel farm in Canton, China, was capable of raising, cleaning, grilling *kabayaki*-style, freezing, and packaging eight thousand tons of eels annually, most of which was shipped to Japan. Farms in China require infusions of hundreds of millions of glass eels annually to keep up with demand.

eel trade remains dependent on the capture of wild fish, because no one has figured out how to reproduce eels in captivity in an economically viable way. In the early 1990s the population of native eels in Japan, *Anguilla japonica*, began to nosedive, and prices for eel became prohibitively high. Asian dealers started to look elsewhere to meet the demand and soon found that similar freshwater eel species lived in Europe and North America. Europe had existing fisheries for eels, mostly for the adults, with several historical eel-fishing centers—the Skåne region of southern Sweden, the Basque regions of France and Spain, Comacchio, Italy, and Lough Neagh, Ireland, to name a few.* Once cherished by Native Americans and early white settlers (one of the first things Squanto taught the Pilgrims of Plymouth Colony was to fish for eels), the eel had been largely ignored as a food fish in modern America.** This made North America, even more than Europe, a perfect target for export to Asia—eels were plentiful and there were no limits on how many could be taken.

---

* Curiously, several of the places where eels are still important culturally, such as the Basque region and Northern Ireland, also support local nationalist resistance groups (ETA and IRA, respectively). You could include the Maori in New Zealand as a nonviolent example.

** On March 22, 1621, the Pilgrims made peace with Massasoit of the Wampanoag Indians. The next day Squanto went fishing for eels to feed the Pilgrims, who were starving. The following account is from *Mourt's Relation*, written contemporaneously: "Friday [the 23rd of March] was a very fair day. Samoset and Squanto still remained with us. Squanto went at noon to fish for eels. At night he came home with as many as he could well lift in one hand; which our people were glad of. They were fat and sweet. He trod them out with his feet; and so caught them with his hands without any other instrument."

One issue that had to be overcome was getting the eels from North America or Europe to Japanese markets alive, since very often they are killed, dressed, and prepared all at once in eel-only restaurants. Instead of shipping adult eels, which was costly (because of the shipping weight), dealers focused on the capture of glass eels, which could be shipped by the millions to warehouse-style farms in China and Taiwan where they are fattened.* Another advantage of farming eels from the juvenile stage is that the fish can be raised to the preferred size, about eighteen inches, so that a single eel, halved, splayed, grilled, and served over a bed of rice, will fit into a black lacquered box with a red interior—a dish called *kabayaki unagi*.**

Initially, the price paid to American fishermen on the docks and riverbanks for glass eels was about $30–$45 a pound. In 1997, record-low catches in Japan sent prices to a high that has not been surpassed since. All of a sudden fishermen were getting paid upward of $250 a pound. Enterprising Asian dealers and American seafood wholesalers began training oystermen, lobstermen, cod fishers, Chesapeake watermen, carpenters, insurance salesmen, and hairdressers to dip or set nets for glass eels. These provisional

---

* The number of glass eels in a pound can vary greatly. Canadian glass eels are small, with about 2,700 fish in a pound. North Carolina eels are about the same; in South Carolina they're bigger, averaging 2,200 per pound. In Maine they average 2,500 per pound.

** Technically, grilled eel without rice is called *kabayaki unagi*, and the dish with rice is called *unagi donburi*, or *unajyu*.

fishermen were happy to work long nights during the glass eel run to put a few extra hundred (or thousand) dollars in their pockets.

Naturally, competition for the best eeling spots in river mouths and estuaries increased. Fishermen began to stake out their territories and defend them with their fists and by packing guns. Dealers were toting tens of thousands of dollars cash to pay fishermen and driving tank trucks full of very valuable live glass eels to airports in Boston and New York to be shipped to Asia. There were incidents of fishermen sabotaging each other's nets, pouring bleach in competitors' tanks, firing warning shots, and getting into nasty fights. A fish that hitherto had been just about worthless now was the most valuable food fish on the coast. With the money came trouble; the eel gold rush was in full force.

At that point, nothing was illegal about the fishery. No one seemed to care about eels in any life stage; no regulations existed to protect them. Glass eels had swum invisibly by everyone for centuries, but all of a sudden they were being exported in staggering numbers, thousands of metric tons, enough to potentially cripple the population. State biologists and conservation officers finally woke up to the mass exploitation of the resource—this kind of fishing pressure was not sustainable. Within a few years the coastal fishery for glass eels and related export businesses was shut down. Today, outside of a limited fishery in South Carolina, Maine is the only state in the United States that allows the export of baby eels.

⌒

Wishing to learn more about the glass eel fishery in Maine, and the international trade of glass eels in general, I contacted a state biologist named Skip Zink, who designed and built ramps and ladders to facilitate the upstream migration of glass eels over Maine's many impoundments. Skip suggested I contact Pat Bryant, who set her nets in the mouth of the Pemaquid River, on Maine's breathtakingly beautiful and rugged granite coastline.

I called Pat at her home in Nobleboro one May day, during midseason for the glass eel run, and asked if I could come up and learn more about the fishery. She said she was very busy, but I was welcome to come by and see her tanks of glass eels and export operation. I had a lot of questions. Did she sell to Asian dealers directly? Did she know of any eel dealers that I might be able to meet and talk to?

"Asian dealers?" Pat laughed in her raspy voice. "Hell, there's one asleep on my couch right now."

When I arrived at Pat's home and business, about five hours' drive from my home, she was just leaving for Portland to meet with her urchin dealer from Japan (eels were her primary but not exclusive export). As promised, there was an Asian man on her couch, a buyer from China with whom she worked closely.

"Jonathan can answer some of your questions," Pat said.

Jonathan Yang spent four or five hours a day "babysitting" the thousands of pounds of glass eels being kept alive in aerated holding tanks in Pat's barn. He guarded them from theft, checked periodically to make sure they were looking healthy, and was simply present in the event that there was a power outage so he could start the generator and keep the aerators going. But mostly he sat on the couch smoking cigarettes.

I sat in a chair opposite Jonathan on the couch. He had a big mop of straw-like black hair, a broad nose, and wide eyes. He wore a black jacket, black pants, and black shoes, and continually scratched his scalp, complaining of blackfly bites.

"You're from China?" I asked.

"No," he said, "I Taiwanese."

"Do you stay here with Pat?"

"No," he said, "I stay in Portland." With his accent it sounded like he said "Poland." He spoke good English but in a choppy and almost comically stereotypical Chinese way, squinting through the haze of cigarette smoke.

I carried in my bag some photos from my recent trip to New Zealand and showed them to Jonathan as a way of demonstrating my interest and travels concerning eels. One in particular caught his interest in an almost devotional way. It was a picture of Stella kneeling on the bank of the small spring-fed creek in Kawhia. Eels were coming out of the water to take dog food off the grass and she was petting them on the head. The eels were very close to Stella's bare feet.

"I know this eel," he said with intensity, "but where, what island? Ah, New Zealand," he said, ashing his cigarette and nodding slowly. He examined the photo even more closely, closing one eye and pointing to the large eels in the photo with an unusually long pinky fingernail. "In Taiwan we call this eel *lo moa.* See how it feeds out of water? It has to be strong to move like this. Our native *lo moa* in Taiwan is almost same, but has spots.* Only *lo moa* can climb up on land to eat. When heavy rains, they come up onto the land. They even climb mountain. But in Taiwan now this eel is very few," he said. "We eat them all."

"You eat big eels in Taiwan?" I asked.

"Yes, only Taiwan people eat this eel. In China no eat. Wives make soup in wintertime from *lo moa*. Make husband strong, make her happy. But it's very expensive. A family or a few wives go to market together, buy one eel, and split in pieces. They cook very slowly for few hours, until it make broth, add ginger root."

The name *lo moa*, Jonathan said, means "like *mafia*, strong and discreet." This term described the characteristics of this particular eel, qualities that wives hoped their husbands would possess once they ate it. Taiwanese men will drink the soup three times a day for two weeks. According to Jonathan, the effect can last up one year. "You can make love every day you feel."**

---

* The large native eel of Taiwan would be the tropical freshwater species, *Anguilla marmorata*, which resembles the New Zealand longfin, *Anguilla dieffenbachii*, but its skin is mottled with marble-like markings.

** One of the greatest stories of the aphrodisiac qualities of eating eel can be found in Brillat-Savarin's *The Philosopher in the Kitchen*. The story, called "The Dish of Eel," tells of a woman known throughout Paris as the "Ace of Spades" who serves a dish of eel to a group of parish priests from the local diocese. After the priests devour the delicious meal, "the reverend men were stirred in an unaccustomed manner, and as a result of the inevitable influence of matter on mind, their conversation took a ribald turn." They told of college escapades and scandalous rumors, but later were "ashamed of the things they had said." They attributed everything to the dish of eel.

Jonathan looked at the photo of Stella and the eels with keen commercial interest, but also with a genuine fascination for their size, their age, and their ability to feed out of water.

"The largest *lo moa* I ever see was in market in Taiwan," Jonathan said. "Seventy pound. But I hear about one, more than one hundred. They ship big eel each in own plastic bag with a single ice block. The skin must stay wet and fish must make it to market alive—otherwise worth nothing." Jonathan said that he had dealt in *lo moa* in the past, but tightening export regulations in New Zealand and their protected status in Taiwan made it difficult to get them.

Some years ago, however, a friend of Jonathan's told him about a unique island he'd visited in Micronesia. On this island there were many freshwater rivers, more than any other island in the region. It rained almost every day, two hours in the afternoon, and then it was sunny and very hot. It was lush and green and beautiful and the people were very nice. And there, Jonathan said, "he find many big eels, just like our Taiwan *lo moa*."*

This place was called Pohnpei (Jonathan pronounced it "Pon-a-pey"), a pristine rain forest paradise—a tiny volcanic island a little more than twelve miles across. It was easy to get there from Taiwan, with a brief layover in Guam. "Make very easy to ship *lo moa*."

---

* *Anguilla marmorata*, the *lo moa* species found in rivers of southern Taiwan, is also native to freshwater streams of the islands of Micronesia and east through Indonesia.

"Have you been there?" I asked Jonathan.

"Yes, few times. When I go there, I fish for *lo moa*. The people see me, they go away. I say, 'Why you scared? Just eel.'"

It was not only the numerous freshwater streams that were responsible for the abundance of eels, but the fact that the indigenous Pohnpeians considered the eels sacred and therefore didn't eat them.

"If you make book about eels," Jonathan said, "you must go Pohnpei. Every village, they have different story about *lo moa*. One story, virgin, she washing clothes in the river. *Lo moa* swim up inside her." He laughed. He said the island was so small you could drive the whole road around its periphery in two hours. There were many pools, too, where the people kept eels and fed them.

Jonathan had once exported eels from Pohnpei. He said that the glass eel business was "much better" and that's why he was focusing on that and nothing else. But then he told me the story of what happened to Mr. Chen, the friend who had first introduced him to Pohnpei, and conceded that this was the real reason he'd stopped dealing in giant eels.

Chen liked Pohnpei so much he lived there for a time arranging several shipments of big eels to Taiwan. One shipment, however, was ill-fated.

Chen had collected about two thousand pounds of eels and prepared them for transport. As usual, he kept them in an open tub of water, a circular tank, outside his bedroom window. The night before the eels' departure to Taipei, one of the fish, the largest, which Jonathan had described as being around ninety pounds, kept Chen awake all night, lifting its head out of the water and crying like a human baby. Chen got a bad feeling and slept little that night, but the next day he went ahead with the shipment anyway.

When the plane stopped in Guam to refuel, Chen discovered to his horror that the eels had been loaded in a cargo hold of the plane that was so cold they had frozen and died. Since the eels were only valuable in markets alive, this was a huge loss.

"Oh no!" I exclaimed, sad at this twist.

"When he got back to Taiwan, Mr. Chen had bad dreams," Jonathan told me, lighting another cigarette. "Every night he dream about eel. An army of eels flying out of the sky, hundreds of eels, hitting him in the chest." Jonathan used both hands to mimic the action of the flying eels coming toward him, and tapped with his fingers on his own chest. "He afraid. He tell me, 'Jonathan, I quit the business.' I say to him, 'Don't tell me those stories. You make me scared.' "

Not one month later Chen died of a heart attack.

As I recorded the details of the story in my notebook I came to the chilling realization that the tales I had heard in New Zealand—about things that happened to people when they took too much from the resource—were very similar to Jonathan's contemporary one. If a Maori were to interpret this story, he or she would most certainly conclude that Chen had ignored the warnings of a guardian eel. The monster eel that cried like a baby was a *taniwha*.

Jonathan asked me why I was studying eels. I told him that my initial interest had come from a fascination for the life history of the eel—the only fish that journeys from freshwater to the middle of an ocean to spawn—and had evolved into an exploration of the eels' importance in world cultures. Jonathan said that he felt we were alike: we both liked travel and to meet interesting people.

Several times during our conversation, Jonathan got a call on his cell phone and began speaking Chinese. The calls reminded me that I had come to Maine to learn about the glass eel trade, yet I could not get away from my fascination for the other reality of eels, what manifested as folklore.

"This is busy time," Jonathan said to me, "peak season for glass eels. We about to make big shipment to China. Over one thousand kilos." I did a quick calculation—five million eels.

Earlier in the year, Jonathan had been buying glass eels, *Anguilla rostrata*, from Puerto Rico, the Dominican Republic, Haiti, and Cuba, when they first entered freshwater there in November. Then he was in Asia in January, buying Japanese glass eels, *Anguilla japonica*, from fishermen in Korea, Japan, and Taiwan.

I imagined a possible scenario for one of the glass eels in Jonathan's next shipment; the extended migration it could unwittingly be a part of. An eel, born in the Atlantic Ocean, netted in the mouth of a river in coastal Maine, is flown from Boston to Hong Kong, raised in a farm of nearby Fujian Province, cleaned, grilled, and packaged in factories on site, and finally flown to a New York airport, ending up on a plate in a sushi restaurant in Manhattan (80 percent of the eel consumed in the United States is pre-prepared in factories in China). That same eel, of course, could end up alive or dead in a restaurant just about anywhere in the world.

Jonathan told me that the majority of eels that Japanese eat are of American or European origin raised in farms in China. They prefer to eat their native eel, *Anguilla japonica*, but because it is rare, it is also much more expensive.

In 1997, Jonathan went to North Korea to buy *Anguilla japonica* glass eels. He couldn't wire money to a bank in North Korea because, he said, the government changes it to local currency and then it's worthless. So he brought a rolling carry-on bag filled with $1.2 million in U.S. hundred-dollar bills. With that money he bought 160 pounds of glass eels, which comes out to about $7,500 per pound. "At that time," Jonathan said, "more than price for gold."

Even in an average year, glass eels are the most expensive food fish in the world.

"This is very big business, very risky," Jonathan said.

The market price for glass eels is based on the market price for adults. But if the price for adult eels falls during the fourteen to eighteen months it takes to raise a glass eel for market, Jonathan's Chinese buyer can go bankrupt.

"One year, the farm sells high, they all drive Mercedes-Benzes," Jonathan said. "Next year, price falls, they're riding bicycles."

Jonathan said he didn't spend the majority of the year anywhere in particular. He was on the road all the time, here and there in rented apartments or small homes—he didn't have a permanent address. "I don't like to own things," he said.

Before he went into eels, Yang traded sandalwood in Tonga, New Caledonia, and Fiji. When the sandalwood was gone he bought two ships and fished giant clams. "In Japan they eat giant clam muscle," he said, "good for sashimi." After losing money with clams he went into the lucrative business of selling shark fins in China for soup. But when he saw dolphins caught accidentally on long-line hooks being dragged aboard ship, beaten to death, and thrown back into the sea, he quit. "When they take the dolphins on the ship," Yang said, "you know, they are weeping—you can see the tears."

He said he did his job because he liked nature, which to me seemed counterintuitive. A man who loves nature shows his affection by exploiting it? He explained that the Chinese believe that the resource can never be used up—"they eat anything," he joked—but he personally knew that it could.

In the meantime, Pat had returned from Portland and the meeting with her urchin dealer. Jonathan had to go visit another eel supplier, Bill Sheldon, in the town of Bath. I gave him a ride because his car was on the fritz. It was a beautiful May day.

chapter six

INTO THE *Ocean*

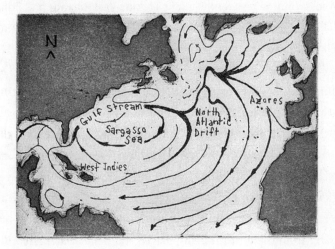

The Sargasso Sea

James McCleave has made more expeditions to the Sargasso Sea to look for spawning eels than anyone else alive. Friedrich Tesch, the German eel biologist, and author of *Der Aal*, now deceased, took McCleave on his first oceanographic cruise in 1974 in the Bay of Biscay. Jim's first Sargasso trip was in 1981, followed by two in 1983, two in 1984, one in 1985, and one in 1989. I met with him at the Inland Fisheries Office in Bangor, not far from where he taught fisheries oceanography at the University of Maine.

McCleave is soft-spoken and modest about his eel work. He is one of the world's top eel scientists, along with Katsumi Tsukamoto of Japan, Don Jellyman of New Zealand, Håkan Wickström of Sweden, Willem Dekker of Holland, and John Casselman of Canada. McCleave and I sat in a classroom around a large table.

"I was looking through a list of my publications over the last thirty years," McCleave laughed, his sea-blue eyes squinting behind his glasses, "and most of them are on eels."

McCleave theorizes that migrating adult eels identify the spawning grounds not by a geographical feature on the bottom, such as a seamount, but by something more subtle—frontal regions, where two different water masses come together at an area of temperature shift. Out in the middle of the Sargasso, water from the north is being driven south and water from the south is being driven north by trade winds, creating many such fronts, sometimes made visible on the surface by big lines of sargassum weed. It is in these areas where Jim and his colleagues have caught the smallest larvae, those most recently hatched, and therefore closest to the adults.

They cruised back and forth across these frontal regions with acoustic gear, hoping to find echoes that would have been of the right strength to come from an aggregation of spawning eels (sonar detects fish underwater by echoing off the air in their swim bladders).

"We occasionally found echoes that we thought might have been groups of eels," Jim said, "but by the time we got the ship turned around and dropped the nets, we were off the mark and never caught anything."

I had a lot of questions for Jim. Where in the water column did the eels reproduce? Were large pelagic fishes out there trying to eat all these eels meeting to lay their eggs? Why did eels go so far to spawn?

Jim smiled politely.

"Dunno," he laughed, and then gave me his scientific best.

"Eels have been around for about two hundred million years," he began, "have survived all kinds of climatic changes, including many ice ages, and have been influenced in geographic distribution and speciation by continental drift. The spawning areas have certainly moved and changed over the millennia, so it's possible that at one time the eels' spawning area may have been closer to the coast."* He reasoned that the American and European eels continued to spawn out in that place, the Sargasso, because it was the only place—with the right temperature and salinity—where the larvae can be reared. "Not all

---

* For instance, when the North American and European continents were closer (and the ocean between them narrower), the eels would not have had to travel so far. But as the plates moved apart and the distance widened, the eels would have had to travel further and further to get to the same general area where optimal spawning conditions exist.

species of *Anguilla* make such long migrations," Jim said, "but all the species that we know anything about spawn in warm, salty water out over deep water."

I asked Jim a series of further questions about eels. After each of them he folded his hands in thought. He twisted his head and then said in his self-effacing way, "We don't know." Then he would amplify: "We thought that . . ." or "We tried to . . ." After years of studying this fascinating fish, not only in the ocean but in their freshwater habitats, Jim acknowledged that in the end maybe it was not so important to know where they went to spawn.

"Part of it," he said, "is everyone wanting to solve the mystery. There's a fellow named Willem Dekker in the Netherlands. He's the chairman of the International Council for the Exploration of the Sea working group on eels. He's done all kinds of population dynamics studies on eels and so on. He's never worked in the ocean, but to him the most burning question is to find the spawning adult European eel in the Sargasso Sea. And I keep saying, 'Willem, you know, what difference does it make if you actually catch one, 'cause we know the little larvae are there?' 'Ah,' he says, 'because we haven't completed the life cycle.'"

But, Jim added, "if we knew where the spawning area was, then maybe we could find out why eel populations are in such severe decline. The declines might be due to some feature in the ocean that is affecting larval transport or larval survival, as opposed to things affecting eel populations on land, like dams and overfishing. But without knowing what's going on in the ocean, it's hard to say."

⌣

Later that afternoon I drove to the Department of Marine Resources to interview Jim's colleague Gail Wippelhauser, who had accompanied Jim on several expeditions to the Sargasso. The expeditions usually left in early February from Miami and involved weeks of towing fine-mesh nets for larvae. She described the Sargasso as a place of "huge thermal gradients and mesocurrents."

Gail spoke of the history of Sargasso expeditions, starting with Johannes Schmidt. After decades of larvae collection, Schmidt concluded in his papers that both American and European eels spawn in a single place, the Sargasso Sea. But, as McCleave points out in his contribution to the book *Eel Biology*, Schmidt's case rested on "a modest amount of published data." The reason the Sargasso proposition has stuck so well is not because it is irrefutable but because Schmidt "presented the case so forcefully that his ideas were largely accepted and still are."[*]

McCleave and others strongly assert that Schmidt's data were not sufficient to support his grand concluding statement in 1935: "In the Sargasso Sea, the newly hatched larvae of our eels were found, and with the aid of numerous sections of fishing experiments we had been able to determine their distribution and settle conclusively that they are found in an area to the

---

[*] K. Aida, K. Tsukamoko, and K. Yamauchi, *Eel Biology* (Tokyo: Springer-Verlag, 2003).

north-east of St. Thomas and south-east of the Bermudas and nowhere else."

The problem with the phrase "and nowhere else" was that Schmidt had barely taken any samples for eel larvae outside of his purported spawning area (in particular south of 20 degrees north latitude). How could he say for sure that the freshwater eel was not spawning elsewhere when he hadn't really looked?

Earlier in his life, in 1922, a more modest Schmidt, younger and presumably less focused on securing his legacy, wrote: "I perceived that if the problem were to be solved in anything like a satisfactory manner it would be necessary to ascertain, not only where the youngest larvae were to be found, but also where they were not. Until a comprehensive survey had been obtained as to the distribution and respective density of the various sizes of larvae in all parts of the sea, it would hardly be possible to form definite conclusions as to the origins of the eels of our European continent."

After Schmidt, the Sargasso Sea was free of eel researchers until 1979, when Friedrich Tesch made his first trip there. This was followed by the various surveys of McCleave from 1981 to 1989, but all these were conducted in the known spawning area, where the anguillid larvae were abundant. And still today, most scientists who are enmeshed in the mystery will tell you that a comprehensive survey of the ocean for eel larvae in other regions outside the known spawning area has not yet been made. The only true answer to the question of whether there is one single or several separate or overlapping spawning areas for American and European eels in the Atlantic is "Dunno."

Gail said that she and Jim used to get together over a few beers and come up with harebrained ideas about how to find spawning eels. "We thought of attaching balloons to migrating eels that would inflate after a certain amount of time and the eel would pop to the surface."

Gail described one method they employed in their search for adult eels. "We brought artificially matured female eels and injected them with hormones to make them reproductively ripe. They put these big females out there in cages as decoys on buoys hoping they'd attract males. We lost sight of them. Things disappear very quickly in the ocean." Gail couldn't hold back. "I was so mad because I injected all those eels. Jim and I made trips to the Darling marine lab, alternating three times a week, a two-and-a-half-to-three-hour drive, to inject hormones to induce sexual maturation. We had a hundred artificially matured females to take out to the Sargasso Sea and lower in cages attached to buoys to attract males. By the time we got to Miami to catch the ship, most of them had died. And by the time we reached the Sargasso, we had only five. We watched the buoys on the radar. Jim and I took shifts following them. And they disappeared. God, was I angry! All that driving."

She laughed, then continued talking about a particular trip to the Sargasso with Steve Brandt, a hydroacoustics expert. "We saw one big aggregation on sonar that we thought might have been the eels. We had five or six nets and never saw one."

Gail spoke of the adult eels returning to the place where they were born, spawning and dying, and then the newly hatched larvae drifting toward the coast from the spawning area. "The currents are so strong, the larvae have no control. They get stuck in the Gulf Stream." Gail offered no explanation as to how the multitudes of larvae dispense themselves to occupy freshwater habitat throughout the extensive range. How did some know to swim up the Mississippi and others the Delaware, the Hudson, or the St. Lawrence? I asked if it was possible for young eels to return to the same rivers, or at least the same general area, that their parents had migrated from.* With no uncertainty at all in her voice Gail said, "No. Young eels don't home to particular rivers. The adults home to where they were born, just like salmon do. Except with eels it happens that their birthplace is an almost featureless place in the open ocean."**

----

* The panmixia hypothesis—that all eels migrate to the Sargasso Sea and randomly mate with each other—was challenged in a paper published in the journal *Nature* in February 2001, "Genetic Evidence Against Panmixia in the European Eel." The authors, Thierry Wirth and Louis Bernatchez, posited that the northern (Baltic Sea) and southern (Mediterranean Sea) populations of European eels are two genetically distinct and therefore reproductively isolated groups. Which meant that when the southern population migrated to the ocean to spawn, they did not mix with the northern population, and that the offspring of each group returned to the same general areas from which their parents migrated. This theory has since been proven wrong. It does in fact appear that the entire European population spawns together and that the offspring are randomly dispersed to freshwater rivers throughout the range.

** The eels' reproductive strategy is very different from other migratory fish that move between freshwater and saltwater. Salmon, because they spawn in rivers and migrate home to the rivers where they were born, have been reproductively isolated in separate river drainages for thousands of years and have evolved into slightly different varieties. Though not significant enough genetically to be considered separate species, each river has a unique population. (In terms of conservation of the species this must be considered. You can't, for instance, introduce Atlantic Salmon from Canada to replace the extinct popula-

Gail said that if she and Jim went back to try to catch adult eels in the Sargasso Sea, they'd bring commercial fishing gear. "We always thought the navy knew where they went," she added. "It's really deep near the Puerto Rican trench. They must test their subs down there. Who has all the latest and greatest toys but the navy?"

Gail and Jim agree that it is just a matter of time before the mystery of the eels' spawning place is unraveled. McCleave writes in *Eel Biology*: "At some time, smaller telemetering devices will allow direct determination of locations and times of spawning of European and American eels migrating from different parts of their continental ranges."

If I live to read that headline in the news, it will be a bittersweet day.

---

tion from the Connecticut River in southern New England and expect them to thrive.) Eels, on the other hand, mix their entire adult population every year reproducing together in the Atlantic Ocean, so the species as a whole is more homogeneous.

chapter seven

WHERE EELS

Go to Die

Japanese Eel Day poster

Tsukiji, which literally means "landfill," is built on a former wetland in the delta of the Sumida River. It is the largest seafood market in the world. Inside is a wonder of the fish trade; outside is a series of restaurants and markets selling everything from sushi knives to T-shirts. The bustle begins outside of Tsukiji well before 4:30 A.M., when we arrived—the planes full of fish are landing at Narita Airport from the farthest corners of the world, trucks coming in and backing out, the auctioning of tuna, the opening of Styrofoam containers gleaming with fresh seafood, dead as well as live, in all imaginable sizes, shapes, and colors.

Both of my travel companions, Kunio Kadowaki (native Japanese fixer/translator) and David Doubilet (photographer), had spent a great deal of time at Tsukiji on various assignments for *National Geographic*. I was looking forward to walking through it for the first time.

The first story that Kunio and David had worked on together was bluefin tuna in 1978. Things had changed in our oceans in thirty years. For one thing, the number and size of our large pelagic fishes, the tuna, swordfish, and marlin, had decreased drastically. Populations of bluefin tuna, an indicator of ocean health, were crashing toward extinction. The major cause of the decline seemed to be overfishing, an increased demand for fish, and a finite supply. Eel populations were not faring much better.

Standing in the muggy morning air thick with the smell of fresh fish, we entered Tsukiji through same door that many a jet-lagged tourist had before. The door leads to a damp, dark hallway that opens onto room upon room of frozen bluefin tuna. Collectively, the tuna, their tails lopped off, white with frost and lined up neatly in rows, make a steaming sea of bodies. Some of them may have come from as far as Australia, the Mediterranean, Nova Scotia, and Cape Cod Bay.

"The king of the market is still tuna," David said, lifting his Leica to take a photo.

At around 5:00 A.M. the tuna auctions begin—men shouting, some holding coring devices used to sample the flesh, bidding, making notes, and dragging them away one by one. In 2001 a single tuna weighing 444 pounds was auctioned off at Tsukiji for $173,000.

Beyond the auction rooms are acres of stalls and alleys with the scent of the sea, where the purchased frozen tuna are halved and quartered with saws and long sharp knives. As the segments of tuna thaw, beautiful pumpkin orange and mango red colors emerge from the flesh, ribbed and clean-lined and still draped in silver skin.

"How can the oceans support all of this?" I asked David, in awe of the abundance.

"They can't," he said.

In the cavernous room where we now stood, narrow aisles led between tanks and crates of fish live and dead. It was like being in the collection rooms of a natural history museum, but without the labels telling you the common and scientific names. Baskets and bundles and boxes of things I could roughly identify as octopuses, clams, mussels, shrimp, salmon, seahorses, lamprey, urchin, squid, cod, eels, snapper, grouper, skate, flounder, swordfish, mackerel, crab, and lobster bulged from all corners. There were of course Japanese characters on signs, posters, and cardboard placards that I imagined described what the creatures were. This world had its own vocabulary, even its own language of commerce.

We came to a counter where three men were cleaning live eels. David's shooting increased in intensity.

—

We had come to document the road's end for the eel. In midsummer, an especially large volume of eels passes through Tsukiji. Eel is the sixth-largest import to Tsukiji overall annually, but now, at high season, it is the third-largest, surpassed in live weight only by tuna and shrimp. More than 130,000 tons of eel are consumed in Japan per year, mostly in the form of a grilled eel dish called *kabayaki unagi*. The Japanese eat more eel in midsummer than any other season because they believe the flesh gives them stamina in the relentless heat—relief from what they call *natsubate*, or summer fatigue.

Eel consumption peaks at the end of July on *doyo unagi*, Eel Day. On that particular day bold, colorful signs, flags, and posters commemorate the event, and eel is sold in every supermarket and roadside convenience store. This year Eel Day fell on the twenty-eighth of July, toward the end of our trip, and the city was appropriately hot.

The custom of eating eel in summer began as a marketing ploy. Actually, the proper name for the day when people eat eel is not Eel Day but *doyo no ushi no hi*, the Day of the Ox. This story is well told in Theodore Bestor's book *Tsukiji*: "An eighteenth-century Edo *unagi* (eel) restaurateur had the bright idea of commissioning a famous calligrapher to make a simple sign proclaiming, 'Today is the day of the Ox.' The fame of the calligrapher ensured that passersby would notice the sign and the eel shop, and make the desired assumption that there was something special about the day and its relationship to *unagi*. Once made, the connection stuck."

Kunio said that such clever marketing was a tradition in Japan. He followed with a contemporary example. "Thirty years ago, they never ate chocolate in Japan. Then some savvy marketing guy started a trend where women buy men chocolate on Valentine's Day. Now chocolate sales are *huge* in February." He added, "It is like you eat turkey on Thanksgiving."*

Eating eel in midsummer may have been popularized by savvy marketing, but there could be some substance to the belief that eel beats summer fatigue. Eel meat has well-known health-giving properties.** It is high in vitamins A and E, containing four times more vitamin A than cheese and eight times more than egg, six times more vitamin E than cheese and three times more than egg. Vitamin A is good for human skin. Vitamin E helps prevent aging. Eel is also rich in fish oils that contain antioxidants to aid the immune system and fight sickness. Because of its high concentration of omega–3 fatty acids, eel has been

---

* In a way he was right. Historians agree that wild turkey was indeed on the table at the first Thanksgiving dinner in Plymouth, Massachusetts, but the meal consisted mostly of seafood, including eel, which is to be expected of a coastal town.

** In Ireland, eel fat has been said to be a cure for rheumatism. In rural Illinois, wearing an eel skin around your waist was known to be a cure for lumbago. The astronomer Montanari believed that an eel's liver facilitated delivery in childbirth.

found to help prevent type 2 diabetes. A native of Kyoto told me, "They have a saying in Kyoto—that the girls have beautiful skin because they eat eel."

Eel is not commonly prepared at home, but is customarily eaten in eel-only restaurants. Part of the reason is the difficulty of pacifying and preparing the unruly fish.* Also, while cleaning an eel you must be careful not to get eel blood in an open wound or your eyes, as the blood contains a neurotoxin, less than one cubic centimeter of which injected into a rabbit causes instant convulsions and death.** For this reason, eel is never served raw as sashimi but cooked or hot-smoked, a process by which the toxin is neutralized. (It is rumored that the Borgia family of Renaissance Italy had a secret poison that they kept hidden in hollow rings and dripped into enemies' drinks during meals, and that the main ingredient of this poison was eel blood.)

In the half-light of Tsukiji I put down my camera and took in the colorful buzz of the place. As I watched the pails and tubs of live eels, their metabolism slowed by crushed ice, I couldn't help wondering if they might have come from a farm in China supplied by Jonathan Yang with glass eels caught in Pat Bryant's nets in the Pemaquid River in Maine. The probability that these eels had been born in the Sargasso and caught as glass eels in

---

* The entry for eel in the famous French food encyclopedia *Larousse Gastronomique* begins with instructions on how to kill an eel: "Stun it by banging its head against a stone."

** See Leon Bertin's book, *Eels: A Biological Study.*

Maine was roughly 40 percent. The rest had mostly come from the Basque region of France and Spain and also had been born in the Sargasso Sea.

⌒

At sixty-three, Kunio was grandfatherly and friendly. He wore a photographer's vest made of black mesh with pouches in which he kept a small notebook, an address book, a camera, and a pen. "I am your guide dog," Kunio liked to say, repositioning his wire-rimmed glasses. "Without me you are blind in Japan."

For the remainder of the day we followed our guide dog around Tokyo, dipping in and out of seemingly hidden districts of the city, behind a wall here, along a river there, through a small park with trees. The city was a dizziness of cars and bicycles and lights and signs that all seemed to melt together in the ferocious heat. Somewhere beyond Akihabara, the hyperkinetic electronics district (picture Times Square multiplied by Las Vegas), we turned a corner onto a quiet narrow street where the shadows of potted Japanese maples and bamboo gave the illusion of relief from the heavy air and frenzied city. A small nondescript door led to a cozy restaurant, Kubota, with six tables.

The owner and purveyor of the 120-year-old eel restaurant, Shoichiro Kubota (his father was eel handler to Emperor Hirohito), met us at the door. The restaurant serves only eel and specializes in *kabayaki*, or grilled eel. Kunio said, "When we talk about a dish of eel in Japan, ninety-nine percent of the time we mean *kabayaki*."

Kunio facilitated our entry into the normally closed kitchen.

Each restaurant has its own closely guarded recipe for the sweet sauce put on grilled eel, the base of which is mirin (a sweet rice wine) and soy sauce.

"Until forty years ago," Kubota said, "we served exclusively Japanese eels, many wild caught from the nearby Edo River— but they became too scarce. Now almost all of the eels we serve are farmed." Kubota did not like to admit that 80 percent of the eels eaten in Japan are raised from European or American glass eels.

"American eels are not as tasty," he said. "Even the French eels are not as good—like American cherries. We like our native things."*

Japanese will pay more than ten times as much for Japanese eel as they will for American or European; the same is true for beef, peaches, and melons. In Mitsukoshi, a 380-year-old department store, individual Japanese-grown melons can sell for over $250. Cows born in Australia are raised the last half year of their life in Japan so they can be called Japanese *wagyu* beef. Recently, Kunio said, there was a story in the paper about how short-necked clams from North Korea had been fraudulently sold for a premium as native Japanese clams. The government

---

* Some of the native eels eaten in Tokyo restaurants were not raised from glass eels but actually caught as adults in the greater Tokyo area. We spent part of a day on the Edo River with an old eel fisherman who fished with traditional bamboo traps—three sections of timber bamboo with the nodes drilled out, bound together with twine. There was no bait involved, no snare; the traps simply worked on the eels' predilection for dark tubular places.

exacted fines. "Most Americans would rather have quantity than quality," Kunio said. "Most Japanese want good product."

⌒

Later in the day, at the Kobayashi Company in another part of Tokyo, I watched eleven men and women at three separate tables cleaning eels, at an average of two and a half per minute. The foreman came around every so often with energy drinks for his workers. At the eel slaughterhouse the pace never lagged. One after the other the eels were taken live out of a tub, pinned by their heads with a steel spike into a hole in a wooden cutting board, and opened; the spine was taken out, and the flesh laid flat into two symmetrical fillets still connected by skin. John Lennon's "Beautiful Boy" played through a weak signal on the radio. Eel spines, flesh with skin, heads, and guts were all placed into separate buckets, the eel heads still moving after being severed from the bodies. In another room three men were easing several bamboo skewers into each fillet and putting them on a conveyor belt, where they were grilled and dipped in *kabayaki* sauce by a machine. The foreman told me they could clean, skewer, and grill an average of four thousand eels a day.

⌒

Leaving the eel slaughterhouse, I headed across town to the Ocean Research Institute (ORI) at the University of Tokyo to meet Professor Katsumi Tsukamoto, the man who discovered the spawning place of the Japanese eel. His laboratory is the only one in the world that is actively doing ocean research on eels.

"We are able to do our research because eels are culturally important in Japan," Mike Miller, who works with Katsumi in the Department of Marine Bioscience, had told me by phone. "The food supports the science. If they didn't eat eel in Japan, I wouldn't be here."

I was fortunate to have caught Mike and Katsumi on land, as ordinarily at this time of year they would be on a summer cruise, sampling the ocean for the larvae of newly hatched eels. Their latest cruise, from December through March, had followed a track from Tokyo around New Zealand to Antarctica and back.

Almost every year the team at ORI makes cruises in the Indo-Pacific region, searching for clues that will tell them more about the life history of catadromous eels. The high costs of operating research vessels (at least $10,000 a day) are assumed by the Japanese government. As Miller said, the research is motivated by Japan's hunger for eel. The thought is that if they knew where and under what conditions eels reproduced, then they might have a better chance of hatching and raising eels in captivity in a cost-effective way. The ability to do so would alleviate pressure on the wild fishery, make eel meat more affordable, and preserve an important part of Japanese cultural heritage.

On entering ORI I passed a young woman in the hall and asked where I might find Mike Miller. Soon enough I was shaking hands with a tall, broad man with a short-cropped beard wearing a plaid shirt; he looked like a lumberjack. Mike brought me into his office and we sat down at his desk. Next to his computer was a framed photo of his expedition team from the previous cruise. He stood beside his much smaller Japanese colleagues, sticking out like a proverbial sore thumb.

Originally from Oklahoma, Mike got involved in eel research through his professor at the University of Maine at Orono, James McCleave. Mike had accompanied Jim on a 1989 cruise to the Sargasso Sea, and though the expedition was not successful in its objective—to catch adult eels on the spawning grounds—he got his first taste of what it was like to be out in the middle of an ocean.

"It's a form of pure nature that most people don't even know exists," he said.

McCleave introduced Miller to Katsumi Tsukamoto at the Ocean Research Institute in Tokyo, and he was invited to join a 1991 cruise in the Pacific on the R/V *Hakuho Maru*. Miller enthusiastically accepted. Tsukamoto was trying to locate the then-unknown spawning area of the Japanese eel, *Anguilla japonica*, by cruising the ocean with fine-mesh nets in search of eel larvae.

It would be the fifth trip for Katsumi and the *Hakuho Maru* in search of *Anguilla japonica*. Only a hundred or so larvae of Japanese eels had ever been captured, and never in high enough concentration or small enough size to indicate the location of spawning. But on that trip in 1991, one evening near the end of the cruise, Tsukamoto, Miller, and the rest of the crew captured for the first time very small larvae of the Japanese eel in the region to the west of the island of Guam and the other Mariana Islands. It was an evening the team came to call the "longest night," as they collected and processed more than nine hundred small eel larvae, a quarter inch long or less, before sunrise.

The discovery of the general spawning area of *Anguilla japonica* was the cover story in the journal *Nature*, as well as *Science News*. It was heralded in the Japanese press as one of the great events of recent Japanese history (and made for an auspicious story, as it happened in July, very near to Eel Day). In part because of the success of that trip, Miller decided to move to Japan to work with Katsumi, specializing in the study of eel larvae, not only the larvae of *Anguilla japonica* but all different species of freshwater and marine eels.

Seated at his desk, Mike handed me a glass vial with about half a dozen preserved specimens of tiny larvae that they'd collected on the previous expedition. It was the first time I had actually seen an eel larva, albeit dead. They were beautiful—like clear willow leaves with a barely visible herringbone skeleton. I swirled the leptocephali in their preservative solution. They momentarily danced on their pin-sized heads before I placed the vial back on Mike's desk and watched them settle like flakes in a snowglobe.

With a combination of perseverance, knowledge (of ocean currents, salinity, and temperature fronts), and luck, Katsumi's team had managed since 1991 to catch larvae only days old, or perhaps just hours from hatching. But despite several attempts with traps, trawls, and various fishing techniques, they still had not captured a mature spawning eel.

By this time in my visit it was late afternoon, and Mike offered me a cold Kirin beer from a small refrigerator in his office. It was very welcome at the end of a sweltering day.

"The ocean is a whole other wilderness," Mike said, taking a swig of his beer. "Sky and clouds and water, with pockets of life here and there. It's beautiful." He tried to evoke the ocean's vastness. "There have been attempts to catch adults," Mike said, "but it's a big ocean. You could be fifty meters away. It's an issue of scale. The open ocean is huge. To get to where eels are spawning is almost impossible. You'd have to be very lucky. You set a trap or pull a net fifty meters on the wrong side, you won't get them."

What made it even more challenging was that the ocean is amorphous and dynamic, and from year to year the spawning area would never be in precisely the same place. Still, the scientists were certain that with the proper resources and persistence the secret would be revealed. I had not realized until I began my conversation with Mike how close they had come. Not all of their latest discoveries had been published—new information about the eel was being gleaned daily, collected not only on ocean cruises but also in the lab. Mike told me that Dr. Hideki Tanaka and his colleagues at the National Research Institute of Aquaculture were making great advances with hatching and raising eels in captivity.

Mike was interested in the larvae of all catadromous eels as well as those of marine eels, but his primary interest these days was to find the spawning area of the longfin eel of New Zealand.

"The longfin is one of the few *Anguilla* species that we have no idea about," Mike said. "We've never identified a longfin larva." His eyes widened at the thought of finding one. "Katsumi funded the pop-up tag research that Don Jellyman had implemented on longfin eels. They got some info," Mike said, "but there were no clear results. It didn't solve the mystery."

After we finished our beers, Mike brought me down the hall to meet Katsumi. He seemed hesitant to interrupt his mentor, but Katsumi welcomed us warmly into his office. He had a gentle demeanor, a tanned face, dark brown eyes, and dark hair with strands of gray. Katsumi showed me the eel icons on votive plaques he'd collected. The eel in Japan, he explained, is among other things a symbol of fertility and safe pregnancy.

Early on in his career, Katsumi became interested in studying fish migration. His first breakthrough was unraveling the mysterious life history of the *ayu*, also called sweetfish, a prized food fish in Japan.* Katsumi's team discovered that *ayu* spawn in freshwater, the larvae go to sea, and then they come back to freshwater in spring (a fish that migrates between fresh and salt water, but not necessarily for purposes of breeding, is called *amphidromous*). "We figured it out," Katsumi said. "But the *ayu* is easy compared to the eel. The eel is very difficult. Takes time to study. We are the closest team. We are very close."

Katsumi had navigated a good portion of the tropical Pacific and visited countless islands in Polynesia, Micronesia, the Philippines, and Indonesia in search of eel larvae. He mentioned that he'd visited one especially interesting island in Micronesia

---

* The traditional way to catch *ayu* is with cormorants. The trained bird dives and catches the fish, storing it in its crop until it is brought to the fisherman.

where there was a clan of people whose totem is the freshwater eel.

"Pohnpei?" I blurted out.

"Yes," he said, "that's it. How did you know?"

This was the island that Jonathan Yang, the eel dealer I met in Maine, had told me about. Katsumi said it was well worth the trouble to get there, a beautiful and wild place with mountains and many freshwater rivers. I made a note about Pohnpei in my notebook.

Katsumi suggested that the three of us go get a bite to eat. But before we left the office he signed a copy of his colorful book of photographs from their voyages on the *Hakuho Maru*, called *Gran Pacifico*. Katsumi said that it included his thoughts about the beauty and diversity of ocean life made during months aboard the research vessel. In the back of the book was a poem he had written about eels. He attempted a translation, reading aloud.

"Why do they migrate so far?" he said. "Why so far, why," he repeated to himself. "Why do they choose this hard life? Why do living creatures live, why do living creatures die?"

Katsumi was in an ebullient mood as we walked down the street away from the office. I felt as though my visit and my questions had caused him to reflect on his travels and his successes.

After two beers at a small bar where Katsumi said he used to go when he was an undergraduate at the University of Tokyo, we walked to a restaurant called Sakuraya to eat dinner. Katsumi spoke poetically about the copulation ritual of manta rays.

"They reproduce belly to belly!" he said in amazement, his face full of childlike wonder. And I thought, *The best scientists are those who never grow up.*

Mike and Katsumi ordered eel and other types of sushi and sashimi. We drank sake and more beer.

Katsumi spoke again of the eels' mysterious migration. He said that long cruises at sea had made him philosophical. "Why spawn, why die?" he said. "Why, why? Eel is very shy and very nervous, but tough, very hard to understand, very powerful . . . a wide range of unpredictable characteristics." He put his head in his hands. "It returns to where it was born," he said. "How does it know?"

That was the last legible thing I wrote in my journal that night. My scrawl had devolved to eel-like scribbles. All I can remember is that there was a warm collective feeling around the table, big smiles, laughter, and a mutual acknowledgment that there were certain things that would never be known.

Why Live? Why Die?

At the end of their long journey the parents spawn
And die
As their children take the ocean currents back
To East Asian rivers from Mariana.
Adults and young both knowingly make their way alone
And through this travel, life is handed down.
For millions of years, birth and death repeats.
It is relentless.
Why do they do these kinds of things?

Where Eels Go to Die

Why do they choose this hard life?
Why do living creatures live?
Why do living creatures die?

—KATSUMI TSUKAMOTO

In Yoshida, an area south of Tokyo known for its green tea cultivation, David, Kunio, and I visited a large industrial fish farm owned by Yoshio Shiraishi, president of the Maruhai-yoshida Eel Culture Cooperative Association. Here Japanese glass eels are fed fish-meal paste in large cement tanks under black plastic tents that keep the fish in the relative darkness they prefer. The 110°F heat in the tents is oppressive but helps maintain the water temperature at an optimal 90°F.

"The hotter it is, the faster the eels grow," said Shiraishi.

The government no longer allows glass eels from outside the country to be raised in Japanese eel farms. Even if they did, conditions could be hostile to American or European glass eels, which require cooler water that is richer in oxygen. Most of the Japanese glass eels are caught in winter and early spring—roughly between December and April. When the wild glass eels arrive at the farm, the first task of the farmer is to acclimatize the juvenile eels to tank life by feeding them a squid-based powder. Shiraishi said it takes six months to two years for an eel to reach *kabayaki* size in the farm. "Some little guys, like human beings, take more years to grow," he said.

The price for Japanese glass eels in 2003 was $900 a pound.

In 2004, it was $1,100 a pound. In 2005, the year I visited, glass eels were $3,600 a pound. The price was so high, Shiraishi said, because "they did not catch many that year." At this farm they currently had 150,000 or so eels in cultivation. Each tank could hold approximately 17,000 adult eels. I asked Shiraishi if American eels tasted the same as Japanese eels. He said if they were raised on the same feed, it might be hard to tell the difference.

Our next stop was the lab that could one day supply this farm with artificially reproduced glass eels.

Kunio did not explain until we reached the gates of the National Research Institute of Aquaculture in Nansei that he had timed our visit to coincide with the day and time—Friday at eleven in the morning—when every week Dr. Hideki Tanaka and his team stripped eggs from ripe female eels and fertilized them.

Tanaka, a slim, soft-spoken Japanese man, met us at the side door to the building and walked us casually around the lab. Room after room was filled with tanks made of clear acrylic in various sizes and shapes, but mostly cylindrical, bathed in a fluorescent purple glow, which, Tanaka explained, simulated the light conditions hundreds of feet below the surface of the ocean. Swimming in the tanks were eel larvae, varying in age from seven to two hundred days. This was likely the closest we would ever come to seeing the early moments of an eel's life in the ocean.

Tanaka encouraged us to peer into the cylinders where hundreds of small clear eel larvae were dancing irregularly. Many had their heads at the bottom of the tanks, apparently feeding on something that had settled to the bottom. The equipment in the room—filters, lights, instruments for maintaining salinity and temperature—were deceptively simple. These eel larvae were the fruit of consistent and rigorous work that had been ongoing since the 1970s.

There was an enormous amount of pressure on the lab to come up with a simple formula for hatching and rearing freshwater eels in captivity. This pressure only increased as wild populations of eels declined. One of Tanaka's papers states the problem clearly: "To maintain the natural glass eel resources and to obtain reliable supplies of glass eels for aquaculture, development of an artificially induced breeding procedure for eels has been eagerly desired."

Eggs of *Anguilla japonica* were first successfully fertilized and hatched in captivity in 1973 at the University of Hokkaido.[*] The newly hatched pre-leptocephali larvae, however, could not be kept alive beyond their first few days. Once the fish had depleted their yolk sacs (stores of nutrients that carry fish through the first period of life), the biologists could not get them to feed on their own.

The difficulty was to find what the cultivated larvae would eat, especially when they had no way of studying what the wild larvae were eating in the ocean. "We tried everything," Tanaka

---

[*] *Anguilla japonica* females were induced into sexual maturity by injections of salmon pituitary hormone, males by injections of human chorionic gonadotropin.

said. A short list of attempted foods included zooplankton, eggs of other fish, rotifers, cuttlefish, shrimp, jellyfish, and mussel gonads. Finally, in 2001, Tanaka and his team found that eel larvae would actively ingest a slurry-type diet made from freeze-dried shark egg powder. On this diet they were able to keep the larvae alive for eighteen days, a record at that point. By mixing the shark egg powder with soybean peptide, krill extract, and a variety of vitamins and minerals, they succeeded in extending their lives further. Finally, when the hatchlings were 250 days old and two inches in length, the scientists watched the leaf-shaped larvae metamorphose into glass eels.

The year we entered Tanaka's lab was the first that they had raised an eel to the size at which they would normally be harvested at an eel farm, roughly eighteen inches. In a tank on the left side of the room two such eels were taking shelter in PVC tubes.* Their heads were misshapen, their pectoral fins abraded. I asked Tanaka if they'd given them names. "No," he said, "just a number."

Tanaka said that his lab requires about $1.8 million a year to sustain all of its projects, two-thirds of which are related to eel reproduction. And though they had had significant success, the ratio of eggs hatched to adults reared was not anywhere close to being profitable in a market environment. After 50 days only 4 percent of hatched eels survived; by the 100th day it was down to 1 percent. A mere handful survived to 250 days.

---

* Three segments of PVC tube lashed together in a pyramid shape, not unlike the bamboo eel traps we'd seen.

The high mortality and slow growth rates in the lab were likely due to a host of factors, but given how little was known of the eel's life cycle in the ocean, they could not be easily determined. Hormones injected into the males and females to make them mature adults caused deformities in the offspring. The eel larvae raised in the lab did not look like larvae caught by Mike Miller and Katsumi in the ocean—their heads were malformed and they seemed to have trouble swimming.[*]

Tanaka took David, Kunio, and me into a small, nondescript white-walled room where they strip the ripe female eels of their eggs. In eel farming almost all of the fish are males. "No one knows why," Tanaka said.[**] To create females for the purposes of propagation in the lab, they feed glass eels estrogen.

---

[*] Some years later I visited a lab in New Zealand (the Mahurangi Technical Institute) where attempts are being made to hatch and raise eels in captivity. I spent a day with the chief scientist, an Iraqi woman named Tagried Kurwie. "We're hatching them by the millions," Tagried said. She wouldn't tell me if they had been able to get the eel larvae to feed, saying coyly, "We believe in a few things that you can get off the shelf." When I mentioned the deformities I'd seen in the larvae in the Japanese lab, a big, knowing smile spread across her face. "We knew there must be flaws in their techniques," she said. "They're doing multiple injections of hormones. We believe the injections are bad for the eels and are producing malformed larvae. If you overdose them on hormone, they will definitely have deformities. One of our breakthroughs is reducing the number of injections from twenty to two."

[**] How an eel becomes either male or female is a mystery, though it may have something to do with water temperature, salinity, food type, food availability, and/or density of the glass eel population in a particular estuary, lake, or river system. Generally in the wild, the eels living upstream in a river system are females, and those occupying the lower reaches and estuaries are males.

Before us was one of three adult females—number 24—that would be stripped of eggs that week. She had just been removed from a bucket, set on a lab table, and anesthetized. She was a sleek eel maybe three feet long, but the entire middle of her body bulged with eggs. The eel gasped slowly, clearly alive. One technician held her while another technician put pressure on her belly, encouraging yellow glutinous masses of eggs to spill out of her anal vent into a beaker.

Tanaka said it was not unusual for one female to produce ten million to twenty million eggs in the wild, but in captivity they are lucky to get one million. The individual eggs are visible to the naked eye, about 1 millimeter in diameter (smaller than the head of a pin), and there are about 2,000 eggs in a gram. I asked Tanaka if he thought adult eels in the wild expire after spawning. He pointed to the female eel stripped of its eggs; she was now dead. "I think it's very much like this," he said. "It is so exhausting for them, they probably die." This particular female was only three years old, while the average spawning adult in the wild was ten to twenty years old. Her eggs were taken to another room, where they were fertilized with milt from a male eel. If, as Kunio had said, Japan is the end of the road for the eel, in Tanaka's lab, at least, it is also oddly a beginning.[*]

* Scientists in the field and in the lab agreed that the current exploitation of wild fish resources cannot be sustained. In the future, if people want to eat fish, they will probably have to make them.

After visiting the lab I traveled south with Kunio to his home city, Kyoto, and settled myself in a hotel from which I could access all the major temples and gardens. Two days later, on July 28, Eel Day, Kunio and I met up for a celebratory late lunch of eel.

"Observe how Japanese people eat eel," Kunio said as we entered an eel restaurant called Hamamatsuya in the Gion district (the Gion is one of the most famous places to see geishas in Japan). We took off our shoes and put on slippers.

Behind a wooden counter the staff were preparing the eels for cooking. One by one the live eels were taken out of plastic tubs and impaled through the head on a metal pin inserted into a hole in a wood cutting board. With the quick sweep of a knife, the eel was butterflied open and the spine and anal and ventral fins taken out, keeping the entire fish still attached by the skin.*

"Japanese food is very delicate—made to order," Kunio noted. "You prepare and then you eat. If you bring it home in a bento box, you eat it soon. If not, it's like yesterday's bread."

After the eel is gutted and the fins taken out, the head is removed and the fish is cut into two equal pieces. The white meat is skewered with four or five bamboo sticks and placed in a basin of water. A few moments later, the skewered eel is

---

* There was one significant difference between the way the eels were cleaned here in Kyoto and the way I'd seen them prepared in the slaughterhouse in Tokyo. In Osaka and Kyoto the eel is cleaned from the ventral side (belly), while in Tokyo the eel is splayed from the dorsal side (back) to avoid mimicking the ritual suicide of the samurai—knife in the belly. Kunio explained that Tokyo was the main city of the samurai and "was more honor-based."

taken out and laid on a metal grill over a hot wood fire. It sizzles and crackles as the fat in the flesh and skin is slowly rendered (several times it is also dipped in water and returned to the fire to steam the meat). The chef rotates a dozen or so eels around the grill surface at once, swapping, turning, and fanning before glazing the meat with the sweet sauce of soy, mirin, and sugar, and sprinkling it with *sansho,* mountain pepper.

As the chef stirred the coals they made a sound like dull wind chimes (Kunio said they use a hardwood from China, like oak). I found it hard to understand how the chef could tolerate the intense heat, but the dry heat from the coals was almost a relief from the muggy, sticky air in the streets of town.

No part of the eel goes to waste. The spine is deep-fried and eaten like a cracker—*hone senbei.* The guts, *unagi kimo,* are skewered and grilled—"No guts, no glory," Kunio said as he sampled a mouthful. The eel liver is served in a soup. Eel is most often served at lunch, at the hottest time of day, as the only course. Kunio said that the Japanese tended not to eat rich, heavy food like eel at dinnertime. You eat it, Kunio says, and you feel like you need to do something, like take a long walk. "Japanese have a taste for fatty food," Kunio said, "but very little. They like a marbled beef like Kobe, but only two or three pieces. And they always eat with rice. It's more balanced."

The eel had a crispy skin and tender flesh. It was rich and flavorful, slightly buttery, and smooth like pâté.*

After my meal with Kunio, I said goodbye and made my way alone through the Gion, imagining that it was spring and the streets were pink with cherry blossoms and peopled by geishas dressed in their kimonos with eel-like trains. I felt a little energetic lift despite the heat; was it the eel I had eaten, or just the elation of being anonymous in a beautiful city?

---

* I had to admit, I liked the taste of eel. I'd eaten Ray Turner's smoked eel, eel fried in New Zealand, eel grilled and then pickled in vinegar in Italy, eel steamed and grilled in Serbia/Montenegro, eel stewed in Sweden. On a spring visit to Maine to see Jonathan Yang, I bought two ounces of glass eels for $60 and brought them home to a Spanish restaurant near my home to be prepared in the traditional Basque way—shocked in boiling water, then sautéed in olive oil with lots of garlic. The baby eels (the Basque call them *angulas*) were served in a ceramic tureen in bubbling hot extra-virgin olive oil, white eels with gray backs simmering with slices of garlic. They are customarily eaten with a wooden fork (to prevent burning your tongue on a metal one). The taste was rich and delicious; the texture was smooth, with a hint of fish flavor. I didn't know what the fuss was about until I tried them. The crash in eel populations worldwide, especially in Europe, is threatening the future of many traditional eel dishes.

chapter eight

EEL

Weir Hollow

Ray at the weir

Summer is a beautiful time at Green Flats and Eel Weir Hollow. The river is low and clear and good for swimming, and the weir is taking shape again after the winter ice and spring floods pulled it down.

At this time of year, Ray's home is cheerful, with a garden of carefully cultivated wildflowers and bushes—black-eyed

Susans, witch hazel, spicebush, hostas, irises, roses, jewelweed, milkweed, jack-in-the-pulpit, and ferns. Stone shelves jutting out from the walls of his house support small pots with herbs. The depleted piles of split wood that heated his home and fueled his smokehouse the previous winter are growing again, the yard is clear of debris from the floods, and the cold storage is filled with smoked eel.

In the warm weather Ray is most often found in his summer uniform, a white T-shirt and cutoff jeans. When the river is low enough he starts to rebuild the stone walls of the weir. Although he builds the walls mostly by himself, he sometimes gets help, especially during a festival he hosts in late July when his friends come down to the river and camp on his property, eat, drink, share stories, and have contests with ancient spear-throwing devices called *atlatl*. Ray figures it takes about 450 man-hours to build the weir every year; out of those, 30 will be contributed by others. At the rate of 8 hours a day, that's roughly 56 days of lifting stones, pushing gravel, and rolling small boulders.

As the stones are being put back into place on the walls, Ray begins to build the wooden trap or rack at the vortex. Initially he needs the assistance of a friend because the studs are long and unruly and setting up the footings in the river requires skillful carpentry (Ray prides himself in his ability to drive a nail underwater). The pieces of the trap or rack—the studs and siding, and the latticework for the floor of the structure—are taken out of storage in Ray's basement and floated up to the site on an improvised catamaran that Ray makes by lashing two canoes together. Upstream of where the trap will be constructed, Ray makes a big pile of rocks, which acts as a cofferdam to divert the flow of the river around his work area. The first pieces of the rack to go in are the six overlapping ramps of wood with one-inch spaces between to let the water through. These latticed ramps start low in the water on the upstream side, building up higher and higher, each overlapping the next by about a foot, forming a long right triangle (as viewed from the side) inclining toward the back or downstream side. The latticed floor is supported by a wall on each side (also partially latticed to let water through) supported at the foundation with more rocks. The finished structure is roughly the size of a small bus—about six feet wide by twenty feet long. It is designed so that in a flood the trap will not be taken downstream, but the high water will flow over the last and highest ramp (to reiterate a Ray maxim, "We're not here to stop the river; we're here to catch eels"). If a flood of this magnitude happens during the run of eels, the eels go right over the top of the trap. Usually during the run, though, the water is flowing over the second or third ramp, leaving the last three or four dry, and the eels are trapped under the overhanging latticework with no way to get out.

Long before Europeans arrived, a stone weir like this one would have been built not by one man but by an entire Indian village, or several, that had set up temporary camps on the river in anticipation of the fall eel migration. In the weeks or months before the run, members of the tribe would be busy repairing the weir, digging out pits where the eels would be held (and sometimes salted), erecting drying racks, and building traps (different from Ray's, more like fish pots) that were designed to fit over the mouth of the weir at its vortex.[*]

⌒

In late summer I started calling Ray to see if he had an inkling as to when the run of eels might happen. I lived three hours from Hancock, so if it appeared the conditions might precipitate a run, I figured I would just jump in my car and hit the road, arriving in time to see the migrating eels fill up Ray's trap. As it turns out, it was not that easy.

Around the middle of September, the remnants of Hurricane Isabel (that hit the Carolina coast with force) brought wind and rain to the Catskills—though not enough to encourage the eels. A subsequent low-pressure system from the west looked like it might drop enough rain to summon the migration, but the rain fell too hard and too fast and before long the East Branch of the Delaware was in flood stage.

---

[*] Fish traps of this kind, working on the same basic principles of structure and form, but sometimes built with different materials, were made to catch migrating fish all around the world. According to historian David R. Wagner, remnants of stone weirs built by Native Americans to trap eels still exist in New England, despite not having been maintained for more than five hundred years. Wagner writes, "Massive group efforts would be required to build, operate, and maintain each weir." Elsdon Best writes in his 1929 book *Fishing Methods and Devices of the Maori* of the weirs (called *pa tuna*) used by the Maori to trap eels: "There were two forms of weirs in former times—the V or double V, and the straight single-fence weir employed in the Whanganui Valley, but so far as I am aware, in no other district."

"There's too much water," Ray told me over the phone. "I have to wait for the water to come down before I can even get out to the rack and reconnoiter." All he could do was wait and hope that the river didn't rise to the point where it would blow out and over the back of his trap, allowing all the eels to go through.

The next morning I drove to Ray's. The rain had stopped, but overnight the river had crested and was high enough that it was nearly covering the whole structure. By afternoon the next day, the river had come down enough that we could paddle up to the weir and inspect it for damage. Despite all the water, the well-engineered rack was in good shape, but in the twenty years Ray had operated the weir, this was the first time he had failed to harvest any eels during the run itself (he had caught a few in the days before the rain). It was a victory for the eels. Ray might catch a few more in the weeks before he removed the rack for the winter, but the majority had gone by.

━

A year later, in early September I called Ray in advance of a storm.

"Is this the week?" I asked Ray.

"Maybe."

"Will the rain bring the eels down?"

"Maybe."

By the time I was able to get up to see Ray, he was bailing six inches of rainwater out of his basement.

A few nights before, the voluminous moisture from Hurricane

Ivan reached the Catskills, dropping eight inches of rain. Flooding was as bad as anyone alive had seen it, forcing the highway department to close Route 17 from Roscoe to Hancock.

"Are you going to check the trap?" I asked Ray. It was Saturday morning, the eighteenth of September. The storm had blown through and the sky was clear, but the river had not yet crested. "The river's going by at thirty miles per hour," Ray said. "No one should be out there."

This was the second time in thirty years that Ray had had water in his basement. "The first time was the hundred-year flood in 1996," Ray said, "and this is worse. I was looking out the window all morning. It was raining so hard, the surface of the river was frosted from the drops. There were trees, picnic benches, refrigerators—all form of human garbage going by. But we were able to save the canoes before it really came up."

Chris Pappas, a slim man with a scruffy beard who grew up with Ray, told me that it was the worst flood he'd seen since 1972, when an ice dam formed above the village and caused the river to back up.

Ray said he was out at the trap until eleven-thirty the night the storm arrived. "I only saw one eel," he said. "The lightning kept them down."

"They don't like lightning?"

"From my experience, no."

If you'd been betting on the eels, you would have won two years in a row. A flood was the eels' best and safest way to salt water. The greater the volume of water, the more efficient it was for migrating fish to overcome obstacles and the more energy they could reserve for their journey to the spawning grounds.

In the two weeks before the river flooded, Ray had caught 650 fish, at least enough to keep him cleaning and smoking for a while.

⌒

In early April of the following year, I drove up to the Catskills to see how Ray had fared over the winter. He was waiting at the end of his muddy and icy driveway for his blue-eyed, pony-tailed, flint-knapping buddy, Chris Pappas, to pick him up and bring him into town to retrieve his truck.

"Drive shaft blew," Ray said. "Very expensive."

Ray seemed happy to see me and gave me a tour of the deck and screened-in porch he'd just added on to the back of his house, furnished with discards from a local nursing home. "Put me in the hole," Ray said, pushing at some exposed insulation. He pulled a wad of dollar bills out of his pocket. "This is all I have left."

Ray offered me a cup of hot chocolate. We sat down at a table next to the stove in his kitchen. His speech was short and clipped; I wondered if it was because he hadn't talked to many people during the long cold months.

"Tough winter," he said, lighting a cigarette. "Dog died. That was hard."

After a moment I asked, "You gonna get a puppy?"

"Naw. Can't do it yet. Come out this summer," Ray said. "June through September I'll be building the rack and the weir."

In late June I was in the Catskills for a friend's wedding in the hamlet of Oliverea. After a beautiful, cool evening, the day had turned pleasantly warm. I decided to steal away from the party and drive over the mountain into Hancock.

When I arrived at Ray's home there was a handwritten note on the smokehouse door to anyone who might happen by. "Come out to the water's edge and wave," it said.

I walked down the path beside the house to Green Flats and the riverbank, took off my shirt, and waded out into the Delaware. I let myself drop into the currents and drifted gently down before standing up on a large flat boulder (Ray called it Duck Rock, and the large eddy behind it was Duck Rock Eddy) to look upstream. The weir was broad and distant, framed by the hills of the magnificent valley. Standing in the water above the weir was a slim figure I recognized as Ray.

Ray doesn't wear a watch, but he has a sundial out by the weir—a nail driven upright on a level piece of the eel rack. Out on the river, though, if you ask, he'll say it's summertime.

He wore a frayed white T-shirt and cutoff jeans. His face and arms were tan. It seemed he was charged directly by the sun. He showed me his fingernails, ground down to nothing from "chinking," or taking handfuls of gravel and piling them up on the inside of the weir. Normally in summer, he said, he worked with his "secret weapon"—an eighteen-year-old named Jaime. "He's worked like a real soldier this year," Ray said.

Even though Ray didn't have the best experience in the army, he had picked up the lingo, using words such as *reconnoiter*, *soldier*, and *vanguard*. He often talked about his time in basic training at Fort Leonard Wood, Missouri, and in Panama "blowing things up"—his fights with the sergeant about overworking the men in 105°F heat without water, his being caught with contraband and being honorably discharged (typical of Ray's Catskill humor, he hung the discharge certificate over his toilet). The engineering he learned in the army he uses in his everyday work. To him, nature is full of engineers—birds building nests, beavers building dams, caddis flies building cases of sand and silk—and he's just another one of them.

"I've been asked, 'Why don't you bring in a dozer?'" Ray said. "That totally goes against the point of the operation. Myself or my friends have touched every one of the stones that go into the wall. That's significant."

Ray paused and looked up at the sun, his shirt and shoulders wet with river water.

"I'm a little out of sorts, Jimmy," he said. "An old friend died yesterday. I was up till six A.M. and slept an hour and I haven't eaten lunch."

I followed him to his canoe, where he had some food in a cooler. He unwrapped a small piece of smoked eel from some tinfoil, tore off a piece for me, and ate a bit himself. It was sweet like honeysuckle and made your fingers glisten with oil.

As the summer sun warmed our shoulders, we sat on the rack and nibbled on smoked eel. I told Ray that back in April I'd attended a hearing in Old Lyme, Connecticut, organized by the Atlantic States Marine Fisheries Commission (ASMFC) to discuss the status of the North American eel population. Ray said he'd gone to a similar ASMFC meeting in April in Narrowsburg, New York, about two hours from his home. More such meetings were taking place up and down the East Coast, Virginia to Maine, prompted by a citizens' petition to list the freshwater eel as an endangered species under the Endangered Species Act (ESA). The petition to list was submitted by two brothers from Massachusetts, Doug and Tim Watts, who had grown concerned about the eels' precipitous decline. Commercial landings for eels were the lowest on record—not just in North America but worldwide.

Ray had more immediate concerns than a decline in the eel population, namely, maintenance on his house, cutting and splitting wood for the smokehouse, fixing his truck, his health, building the walls of the weir and the rack for the fall season, and hoping that a flood would not ruin his catch for the third year in a row. There were easier ways to make a living (I almost felt like he would continue to operate the weir whether the eels showed up or not).

That day I helped Ray shift, tilt, and tumble some of the larger stones into place, using an iron bar as a lever. A few dead shad, summer casualties of the rack, littered the bottom like silver coins in a wishing well.* Two bald eagles sat patiently in the trees above us, waiting for the shad that they would take as soon

---

* By law, Ray had to throw the shad back into the river to return to the biomass. Oftentimes he buried them in the wall itself; whether that was functional (the smell may attract eels) or spiritual, I don't know.

as we cleared the area. At some point it occurred to me that I'd better return to my friends' wedding in Oliverea.

⌒

Late that summer Ray spoke confidently about the work he'd done on the weir. Cosmetically and functionally, it was the finest one he'd ever built.

"The weir is at a place where it's close to done," he told me over the phone. "The anxiety is gone. Now I can just go up there to admire and tinker."

The large flat "paving stones" that sloped on the outside of the weir's walls were seamlessly fitted. The pebble-sized gravel on the inside of the walls, the "chinking," was in place and had been further sealed by leaves of Japanese knotweed that floated downstream—torn off their stalks by beavers. "Nature's black plastic," Ray called it. The curve of each of the two convex walls was the best he'd ever created, forming a braid of currents at the vortex of the trap with just the right pull to coax the eels.*

He had made improvements to the wooden rack as well. The eels would still be caught underneath the overlapping lips of each slatted ramp, but in the pockets where he would normally have stood all night, using a dip net to scoop the captured eels into a canoe, he put escape paths of PVC piping that led the eels into a big cage, a holding pen with a top. He could now empty

---

* Over the years, Ray has learned through experience that straight walls are not as efficient as curved walls.

each night's catch the morning after the run, in the light of day, while relieving stress on the rack and keeping predators such as eagles, herons, osprey, and bears from stealing his fish when he wasn't looking.

The beautiful condition of the weir and Ray's pride in his creation seemed to promise a fruitful catch. Now the weather needed to cooperate.

⟜

In late August a large tropical storm began to form in the Caribbean and eventually morphed into a full-blown hurricane. I called Ray on the morning of August 29, and he answered with his customary greeting: "Smokehouse."

"Hey, Ray, it's James."

"What can I do for you, Jimmy?"

"Well, I see there's a storm coming, wanted to check in."

"There's a band of rain from central Pennsylvania to the Yucatan," Ray said. "Based on my experience, I'd have my boots on for tomorrow night."

I got in my car and headed to Hancock.

The weather forecasters were noncommittal about the path of the storm (now a category 4 hurricane called Katrina), but the system was carrying a lot of water—enough to trigger eel runs on every major East Coast river, in domino-like fashion, from south to north. A week before the first drop of rain fell, before the hurricane had even touched the Gulf Coast, the eels in the Delaware River had begun to move (the early eels Ray called vanguard eels). By August 29, Ray had caught 926 eels, more than he'd ever taken by that date. Only the year 2000 had been close, with 840.

"Just look at the numbers," Ray said, one eye on his logbook, one eye on the Weather Channel in his living room. "That storm is hundreds of miles away, and they *know* it's coming."

It had begun to rain, and all through the night the rain continued, but by dawn the river had risen only slightly. The land was very dry and thirsty for the water, and the reservoir upstream was below capacity.

We spent the day preparing—cleaning out the canoes, filling buckets with salt, draining the eel tank and filling it with fresh water.

Toward nightfall we returned to Ray's house for a dinner of "pancakes"—oatmeal and corn, flour and Bisquick and raisins. It was pretty tasty. Ray offered me a beer while he himself drank out of a large mason jar, a 50–50 mix of orange juice and water with a can of ginger ale. After dinner he sat in his chair by the fire and clipped his toenails, a lit cigarette in his mouth. Ray placed a tray on his lap and rolled two cigarettes for our trip upstream.

We walked down the riverbank to where he had his canoes tied up. We each got in one and paddled upstream. It was a warm, dark night—the new moon was only three days away— and a low mist hung over the water. The air was filled with the damp smell of early autumn, though in daylight the trees had only hints of turning color. The sound of the current was still gentle, but the rain continued to fall lightly, teasing the hope that at some point the river would rise, but not rise too much.

We walked along the walls of the weir, our eyes adjusted enough to see the beauty of his workmanship. Near the base of the rack the walls were ten feet thick. He tapped the rim of his baseball cap in homage to Jaime, the young man who had helped him most that summer. "Jaime did this in a truly soldier fashion," Ray said, pointing out a particular feature of the wall.

We climbed onto the damp and slippery slatted ramps where the eels are actually trapped. Looking upstream, I could see clearly the two arms of the weir and the smooth surface of the river above the V. The river made a soft thundering sound as the braid of water funneled through what Ray called the "slip-drop" into the rack and out beneath under our feet.

I helped Ray clear the ramps of debris—sticks and knotweed leaves—and then we sat at the top of the rack under a rain fly he'd rigged to stay dry while we waited for the eels. Ray took one of the cigarettes he'd been keeping under his hat, lit it, took a puff, and stared off at the mountain.

"You can stay here till midnight," Ray said, "and you've had the whole experience."

He passed me the lit cigarette—it was shaped like a snake that had swallowed a mouse. I was mesmerized by the sound of the rain on the tarp over our heads and the droning of the waterfall created by the rack.

"I've been to heaven," Ray said, "and this is it." He'd once told me that his ideal death would be to drown in the river.

"I wonder where they're going," I said after a pause.

"They're going someplace," Ray said.

A warm breeze blew through the rack.

Ray took another puff. "You save your hair when you get it cut?" he asked.

"No," I said.

Some time passed. The drizzle pattered on the rain fly over our heads.

"I save mine," he said. "I haven't cut my hair since 1969. Whenever I brush my hair, I save what comes out. I also save my fingernail and toenail clippings. I don't want any of my DNA floating around out there. 'Cause when I go, it all goes in the box with me."

This paranoia about hair seemed odd yet familiar. Ray's habits, I realized, hewed closely to the rituals I had learned from Stella and her sister, Wiki, that Maori customarily never cut their hair at night for fear that a spirit might take it. And when they do cut their hair or their fingernails they bury it so that no one can get hold of it and put a curse or *makutu* on them. The fact that Ray saved his hair was not so much weird to me as fundamental among people of nature-based faiths.

Talking about the weir, Ray said, "Isn't that a sublime effigy to Mother Nature? Last night there was a mist in front of that hill and it looked just like an elderly pubic mound. Look at the walls of the weir. They are the legs of Mother Nature, and we are sitting in pleasure central. Tell me that isn't true. Nowhere is nature so motherly portrayed than in this weir. We are in the structure of fertility. Think about it—it's a very feminist thing. I feel it so strongly."

I felt it as well. It was uncanny to think, too, how this graphic representation of Mother Nature was made more effectual when the eels swam into it, like thousands of spermatozoa.

"This night she's in good shape," Ray said. "Look at the water coming over that slip-drop. That is a perfect formation. In the dark she looks young tonight."

In the hours until midnight, we didn't see any eels come into the trap. I spent the night at my friend's cabin across the river in Peas Eddy, and was back at Ray's house in the morning.

At about eight o'clock I found the eel fisherman at the start of his routine.

"Make coffee, feed the emu, check the eels in the tank, have a cigarette, take a shit, watch the Weather Channel," as he himself described it.

We spent the whole day on the river. It rained all morning and then the rain let up. Before our eyes, the clear river began to rise and get color—a milky bluish green as an overture to muddy brown. Ray was nearly certain now that the run was going to happen that night, or at least a good preliminary push of fish.

Ray and I went upstream after dark to check the weir. There was noticeably more water coming through the slip-drop into the rack at about nine o'clock than there had been the night before at the same time. We stood in the wooden structure of the rack, the rising water thundering beneath our feet.

In the foamy waterfall, glistening in the darkness, where the river met the ramps of the rack, I suddenly saw a shimmering black and silver fish shimmy up the wooden slats and tumble into the trap. Another came over the ramp, then another and another. It was what I'd come here to see; it was happening— the run. I started down from the top of the rack to the slippery part where the eels were coming over. Before I could get to the waterfall, Ray urged me to turn away.

"Let's go back, kid," he said to me.

I was surprised and torn. This was Ray's weir; how could I stand out here without him? But I'd come to see this wonder of nature. Over the years I'd imagined what it would be like, a surge of life down the river. I stopped moving toward the water as more eels came over the top of the slats, unable to beat the force of the falls and the shelf of wood over their heads to get back.

"I'll call Jaime in the morning to come down and help transfer the eels."

"You don't want to stay out?" I asked, needing to say something.

"There's no reason to. They'll all be in the holding cage in the morning. You're going to need sleep. Tomorrow we'll be bringing eels down to the house for half a day, carrying them bucket after bucket by the hundreds up the hill from Green Flats to the tank. And the tank, as it is, is overflowing with fish. I'm going to have to rig up another tub with an aerator and then hope the power doesn't go out. And if there's still time, we'll start processing these eels to make room for more."

I felt powerless. Was it all business, a military operation? Was it more efficient to leave the weir so we would be well rested for the work the next day? Or was it something else? I couldn't help but wonder if Ray wanted the brunt of the run to remain veiled, and whether it had to do with the sanctity of what he called Mother Nature. Out of respect, I did not question his decision further.

⌒

In the four years that young Jaime Galietta had worked with Ray, this was the first run that he'd witnessed. The previous two seasons the trap had been blown out and the times before that he'd been in school. Ray was generous and social, not simply the odd recluse I had first imagined him to be. Of all his friends, Ray talked most about Jaime, but it was the first time I'd met him. He was a modest and quiet strong young man with a shaved head and a handsome appearance.

"What qualifies it as a run?" I asked Ray that morning.

"If it's close to a thousand," he said, "that's a run."

We made our way up to the weir with anticipation. Ray and Jaime paddled up in one canoe and I walked up the riverbank, wading out in the high water to meet them at the trap. After tying the canoe to the rack, Ray climbed up onto the wooden structure. Jaime and I followed.

"See that?" he said, running his hand over the wooden latticework, water gushing between the slats. "That's eel slime." He then pointed to scratch marks visible against the wet wood. "And that's where Baldy tried to get at the eels."

The talon marks on the wood of the rack had been made that morning. Upstream, one massive white-headed perpetrator was sitting in a tree looking at us. "He can see these fish," Ray said, "probably got a few. Their vision is incredible."

Ray opened the top of the cage and sank his dip net to the bottom. Jaime and I steadied the canoe, watching as Ray ran the head of the net this way and that. Finally he pulled up the net by the handle. It was loaded with beautiful silver eels.

"There's a lot of them, boys!" Ray said, shouting over the rushing water, dumping them into the boat. "Take a look. The bottom of the cage is black with their backs."

Ray kept loading the canoe with fish until the bottom was covered. The eels banged and clanked against the hull, snaking up the gunwales, unable to get out.

The three of us, wading along the shoreline, carefully guided the canoeload of eels downstream to Green Flats. When we got back, Ray gave orders. He would stay with the canoe at the riverbank and unload the eels into buckets. I would carry the buckets up to the smokehouse, and Jaime would count the eels, one by one, as he dumped them into the tub.

It took several trips to the weir and most of the morning to get all the eels up to the smokehouse. Seven hundred forty-nine fish had tumbled into the trap in the previous ten hours or so, more in one night than the totals for each of the previous two years. So far that year, Ray had caught 1,758 eels, all of which were alive in the wooden tank outside the smokehouse, the slats dripping water and covered with moss.*

---

* The tank, like a big barrel cut in half, Ray told me, was from an old trout hatchery, made by a silo company in Unadilla, New York.

"We should get as many eels again tonight," Ray said. He was afraid he'd run out of space to hold them, so for the rest of the afternoon Ray and Jaime began the work of processing the eels to make room for more.

They put sixty to eighty live fish into a vat of salt for forty-five minutes, where they essentially suffocate and die. Then they are turned in an old cement mixer with gravel, which takes the slime off the skin as they tumble.* Once deslimed, the eels are cut open along the belly with a knife and a spoon is run along the spine to remove the kidney and any remaining internal organs. The cleaned eels, their skin almost a translucent violet, are hung by the head with strings in a walk-in freezer. Ray says that freezing them "crushes the cells," allowing excess oils in the fish to drip out while smoking. After being frozen, they are brined for two days in a mixture of salt, brown sugar, and dark fall honey, and finally smoked at 160–180°F in a cement chamber connected via pipe to a barrel stove burning with hand-split applewood.

Ray and Jaime cleaned eels side by side in the smokehouse.

"Of all the people who've helped me with processing, no one cuts them open faster than Jaime," Ray said.

Jaime gave a quiet smile.

After a pause Ray asked him, "Who won the match today?"

"What match?" Jaime asked.

---

* This is an innovation unique to Ray. Others use ashes or detergent to get the slime off the skin. The Japanese leave the skin on the eel and don't bother to deslime. Still others skin the eel before cooking.

"U.S. Open," Ray said. "Tennis."

And then the men were silent for a while except for the clink and clank of their focused work.

～

Time and again I visited Ray. Initially I had wanted to see the run, but after that objective came and went I found myself returning to Green Flats anyway. Maybe it wasn't the run I was interested in so much as Ray. Or maybe he *was* the run—the persistence to live life, to rebuild the walls, in the face of life's hurdles. No matter; I felt sometimes as though there was nothing to learn or figure out, really, just stuff to take in. I told Ray once that I admired the way he lived. "Don't look up to me, kid," he said, "look with me." Nowhere else did his homespun aphorisms make more sense than when I stood on the river beside him and heard the words come out of his mouth as he was lifting those stones: "It's not the journey, it's the road," or "Art is nature out of proportion."

I helped Ray with some odds and ends one day in June, and toward evening I drove into town to get us a pizza. Ray was appreciative of the treat. He offered me a beer.

"Why'd you quit drinking?" I asked him.[*]

"Let's see," Ray said. "Was it the time we were in the toboggan being towed behind a truck in the snow and we went off the road and hit a tree? Or all the times I crashed my car in hit-and-

---

[*] I never asked him such questions about his life until years after I first met him.

runs? Or was it the time I got up in the middle of the night, not knowing where I was, and fell out of my bed—mind you, my bed is ten feet off the ground—and nearly gouged my eye out on the corner of the dresser? I haven't had a drink in nineteen years."

⌒

One day I ran into a childhood friend of Ray's named Ken Mason, who lived up Peas Eddy Road. Ken had been described to me as "a thinker who went up every day into the woods to read books." I told Ken I'd been visiting with Ray, and he shared some insights into Ray's life.

"My sister Janet lived at Green Flats with Ray in the late seventies for three years," Ken said. "She tended the garden and lived barefoot. Hippies, you know."

I asked Ken why he thought Ray lived out there alone.

"The beavers come up into the oxbow near Peas Eddy," Ken said. "They build their dams every year, and every year the ice and spring floods scour them out." I saw where Ken was going with his metaphor.

"Why does he keep rebuilding the weir, year after year?" I asked Ken. "What gives him the energy?"

"Pure and simple," Ken said. "Freedom."

chapter nine

THE LASIALAP OF

Pohnpei

Pohnpei is a five-million-year-old volcanic island cloaked in rain forest, ringed by a barrier reef and a blue, blue sea. It is thirteen miles across and 2,500 feet in elevation, high enough to have cloud forest habitat, supporting thirteen endemic birds and more than a hundred endemic plants. It is as wet as anyplace on earth, with an annual average rainfall of over four hundred inches and more rivers than any other island in the region. It is tropical and hot.

Ever since I had heard Jonathan Yang's story about the giant eels of the island and the failed shipment of live eels to Taiwan that had doomed his friend Mr. Chen, I had wanted to go there. But I felt that without having someone who could be my way in, as Stella had been in New Zealand, it would be a waste of time. Then I met that person unexpectedly at an international conference of the Nature Conservancy in Quito, Ecuador. He was giving a presentation on the disintegration of the reefs and rain forests of Micronesia. I asked him if he had ever been to Pohnpei. He knew it well enough, he said—he had lived there more than half his life. He spoke the language fluently and was married to the daughter of a high chief.

Bill Raynor, fair-skinned and tall, originally from Lodi, California, confessed that his knowledge of eels was limited, but confirmed that the eel was very important to Pohnpeians, especially the Lasialap people of U (pronounced "ooh") municipality, for whom the freshwater eel was a kind of totem. His area of expertise was the indigenous flora of Pohnpei and traditional methods of agroforestry. But even in his work on island plants he could not escape the eels' significance.

"Eels are part of the hydrology of the island," Bill said. "The Pohnpeians believe that the eels help keep the rivers open and free-moving—if you take the eels out of the rivers, the water will stop flowing."

If I traveled there, Bill said, he would make introductions that would facilitate my research.

⌐

After a twelve-hour layover in Hawaii I was happy when the plane finally left for Micronesia. We landed on a handful of islands—Majuro, Kwajalein, Kosrae—before finally making our descent over the outer atolls to Pohnpei.

Bill was waiting at the Kolonia airport to pick me up when I landed and took me in his beat-up four-wheel-drive vehicle into town. He was wearing shorts topped by a T-shirt with holes. He advised me to surrender to the heat and rain.

"Be prepared to get wet, muddy, sweaty, and dirty."

He dropped me off at a small hotel in Kolonia called Yvonne's. Bunches of bananas hung on hooks outside the first-floor rooms for guests to eat. From Yvonne's, Bill said, I could walk everywhere in town, and in the open-air lobby there was always an assortment of expatriate guests milling around.

⌐

Intermittent squalls were followed by blasts of sun. When the rains came, it was as if someone had opened the floodgates and then shut them just as abruptly. In the wake of the showers there was often a rainbow, sometimes two simultaneously. The endemic maroon parrots, Pohnpei lorikeets, with hints of green and yellow on their wings, could be seen flying in chattering flocks over the causeway to the airport. From there also was a clear view of Sokehs Rock, a massive basalt outcropping that is the signature geological feature of the island. Every morning at first light, colorful reef fishes were unloaded at a street market along the sea. The diversity of fishes was almost unbelievable— blue parrotfish, pink and orange snappers, wrasses in rainbow reds, yellows, and greens.

One of Yvonne's permanent residents was an American named Karen Nelson from Wisconsin who taught English at the College of Micronesia (COM, a community college—the closest university was in Guam). I stopped to talk to her in the lobby while getting my morning coffee. She looked very tired—white skin with fig-colored shadows under her eyes— as she waited for a taxi to take her to the school. Oftentimes the taxi was very late.

"If there's anyplace in the middle of nowhere, it's here," she said of Pohnpei. "There's no tourism, in part because Continental Airlines has a monopoly on flights and they can charge whatever they want. And it takes a full two days to get here from just about anywhere." She said her Pohnpeian students liked it that way—the thought of tourists walking through their sacred sites unnerved them.

Karen's taxi arrived, and as she left, a mustachioed man in a kind of guayabera shirt came into the lobby and joined me for a cup of coffee. We sat on multicolored couches and chairs, watching the rain. He was a diplomat from the nearby island of Kosrae. He asked where I was from, and watched me making notes and sketching the view out of the lobby toward the sea. Seeing I was new to the island, he offered his views on Micronesian culture.

In Micronesia, he said, self-worth is based not on an individual's achievements but on modesty, humility, and knowing that one has a secure place in the family. Attempts to get ahead are frowned upon. "The tallest tree is struck by lightning" is a popular saying. Instead of building a home bigger than the neighbors', people endeavor to build their houses smaller.

Such modesty, the Kosraean cautioned, can be a veil for laziness. The social position about not wanting to stick out had become an excuse to do nothing. And the already languid attitude of the people was exaggerated not just by the incessant heat and humidity but by a popular narcotic drink called *sakau*, known elsewhere in the Pacific as kava. Most Pohnpeians viewed *sakau* as one of the island's virtues (there is no dispute that cannot be settled over drinking *sakau*), while others felt it was impeding Pohnpei's progress (it dulls the central nervous system).

The Kosraean grabbed the coffeepot and refilled our cups. He got a twinkle in his eye and leaned forward.

"I heard you talking about eel," he whispered. "You know, on Kosrae we *eat* eel."

"You eat—" I started.

"Shhh," he said. "I don't want to offend the girl working at the front desk. She is Lasialap, the eel clan."

He leaned back, stirring sugar in his coffee.

"Personally, I love eel," he said, "but to her it would be like eating a human. The Lasialap believe eels are people, you know."

I asked him how eels are prepared in Kosrae. He licked his lips.

"What we do in Kosrae is, we put the eel in boiling water to take off the slime and then cook it in an *um*, or earth oven. We wrap the eel in a special leaf that seasons the meat. You don't want to stop eating—it tastes really good."

⌒

Kolonia is Pohnpei's biggest city, though it is very small. They have no traffic problems, so there are no traffic lights. You would be hard pressed to find a stop sign. Storefronts are eclectic. You might see a shop offering "gold teeth services" (dentistry) next to a Mormon mission. Other shops sell goods carved from ivory nuts (the seed of an endemic palm), bracelets made of hawksbill turtle shells (which can't legally leave the island), hardware, or photocopy services. There are a few places to stay and a few restaurants, representing cuisine from cultures that have occupied the island—German, Japanese, American.

The Visitors' Bureau is on the eighteen-acre campus of the Agriculture Station, a series of bunker-like buildings shaded by giant breadfruit trees. In front of the Visitors' Bureau is a rusty sign with arrows indicating the distance to major cities of the world: New York, 8,158 miles; Paris, 10,326; Cape Town, 10,187; Melbourne, 5,288.

Inside, a couple was bargaining with a man selling black pearls. On the back wall was an array of posters promoting everything from dental hygiene (the habit of chewing betel nut, *Areca catechu*, had destroyed many Pohnpeians' teeth) to the consumption of native carotenoid-rich foods—bananas, taro, breadfruit, and pandanus.*

I asked a woman mopping the floor where I could find Edgar. She went behind a door and moments later a man appeared, dark-skinned with dark hair and wearing a light blue button-down shirt and long gray pants.

"Bill told me you might be able to help me," I said. "I'm doing research on—"

"Oh yes," Edgar said, fidgeting with his glasses. "Bill told me about you. I'm very busy at the moment. I do know an eel story, but not very well. There are people who know it better. You need to visit the high chief of the Lasialap. His name is Sahngoro. His Christian name is Elter John. He is also the leader of U—the *nahnmwarki*."

"Can you introduce me to him?" I asked.

"Well, it's not that easy," Edgar said. "You see, someone has to bring you to him formally, someone with a title, and there has to be a *sakau* ceremony. I could not bring you to the *nahnmwarki* because I am not a chief and I am not Lasialap. And someone would have to teach you the proper etiquette. For instance, your head can never be higher than the head of a *nahnmwarki*. I think Bill has already asked Adelino Lorens, our deacon and minister of agriculture, if he could help. Adelino is high up in the Lasialap clan."

---

* The replacement of indigenous foods with those from the West (soda, processed meats, white sugar, white flour, white rice), the poster said, had created a public health disaster on the island: vitamin A deficiencies, type 2 diabetes, heart disease, and cancer.

"Can you tell me your eel story?" I asked Edgar. He smiled, taken aback at my directness.

"It is the story of how the eel came to Pohnpei," he said. "In this story, the starling, called *sloahk*, brought a seed from another island, and that seed became the first eel, which we call *kemisik*. But for now, I'm afraid, that's the best I can do. I'm late to a meeting.

"Some advice," he added as he walked away. "There are not many keys in Pohnpei. But *sakau* is the key to unlock everything you're looking for."

I stopped to look at a poster about the forests and birds of Micronesia. There, among the cardinal honeyeater, native pigeon, kingfisher, and fantail, was the Micronesian starling, the bird that had brought the eel to Pohnpei. It was black, with yellow eyes.

In the early afternoon, Bill met me at Yvonne's and we walked toward the sea to the office of the Conservation Society of Pohnpei (CSP), a nongovernmental organization he helped found in 1998. CSP was the first platform for environmental conservation on the island, and one of the first in Micronesia.

One of CSP's main objectives was to prevent the further deforestation of the island. Cultivation of sakau for commercial export—to Guam, Saipan, Hawaii, and Kansas City (where there is a large Pohnpeian diaspora)—had set off an ecological imbalance that had never occurred when crops were grown on a small scale strictly for domestic consumption. Since the mid-1980s, clearing of trees (mostly for *sakau* cultivation) had reduced the native forests of Pohnpei by more than two-thirds, to only 15 percent of the island's area.

To combat the devastation, CSP staff initiated the Grow Low Sakau campaign, encouraging farmers to cultivate the lowlands, where the forests had already been degraded. The challenge is that farmers prefer to plant their *sakau* in the uplands, where the soil is richer and their crops grow larger and faster. But by clearing hundreds of acres of steep forest hillsides, they were causing erosion and landslides that destroyed their own crops. Without the trees at the headwaters of streams and their shade and roots that retain moisture, the streams were drying up— which was having an ill effect on the freshwater ecosystems, including the population of native eels.[*]

In the CSP offices, with stunning views of the lagoon and outer atolls, Bill called the dozen or so young staff members together.

"This is James," Bill said, addressing the circle of people around the room. "He's visiting from the States to learn more about eels in Pohnpei. I think his time here is a good opportunity for us to consider our freshwater ecosystems, which we

---

[*] As Bill Raynor wrote in a paper in the *Journal of Micronesian Studies*, "The upland forest serves several important ecological functions. Perhaps most important is that the extensive root system and the litter layer of forest vegetation serve to capture rainfall, retard surface runoff, and improve the infiltration of the water into the soil where it is then filtered and slowly released into the streams and rivers that eventually make their way to the coastal mangroves and the lagoons."

don't really know a whole lot about." Bill continued on, switching from English to Pohnpeian. As he spoke, the staff members nodded, looked at me, and smiled.

"Okay," Nixon, one of the CSP employees, said at last. "Marciano and I are going to take you to Pwodoi, a village in the municipality of Kitti," pronounced "Kichy." "In the creek there is a pool where the local children feed the eels and swim with them."

The single ring road that follows the coast around the island (completed in 1986) winds through the lush forest with intermittent and shockingly beautiful vistas of the Pacific Ocean. Because it is the main avenue for foot and car traffic and because of the road's poor condition, it can take a good long time to get from Kolonia to anywhere else. But no one seems to be in much of a hurry.

A few miles east from town, Nixon pulled off the road and we walked down a path through the trees. In the distance I could hear a kind of percussive music, a two-tone *pank penk pank penk*. Eventually we came to a clearing. A rooster and some hens picked in the soil around the smoking embers of a fire. A blackened kettle boiled, steam coming out of the spout. Beneath a thatched roof held up by the hard trunks of tree ferns, two men sat at a large flat slab of lava rock. Shirtless, baring their strong deep-brown torsos, they worked rounded stones over the slab, producing the hollow and metallic sound I'd heard from a distance. They were pounding the roots of *Piper methysticum*, otherwise known as *sakau*.

The root, a beautiful Medusa-like tangle, was dug from the ground, scrubbed of soil with water and brushes, and cut into pieces (some *sakau* root clusters weigh as much as five hundred pounds). The pieces were placed on the slab of lava rock, called a *peitehl*, and mashed with round river stones. Water was added slowly to make a soggy pulp. When the root was sufficiently macerated, it was laid on ribbons of slimy bark, freshly stripped from the trunks of hibiscus bushes. One man gathered and twisted the fibers of the hibiscus, encircling the crushed root, and wrung the bulge like a wet towel, while the other man captured the fluid that streamed out in the half shell of a coconut. As the *sakau* was squeezed, the slime of the hibiscus bark emulsified the oils in the mashed root (which contain the narcotic agent), suspending them evenly throughout the drink.

The first full cup of *sakau* was handed to Nixon. He took a drink, then rotated the cup a half turn and handed it to Marciano. Marciano drank and turned the cup again, handing it to me.

"You're supposed to close your eyes when you drink," he said, "probably because it doesn't look very appetizing."

It slid and slumped down my throat, mucus-like and slimy. The cup was refilled and passed around again. Before we left, Nixon gave the man a few dollars. We got back in the car and headed along the road to Kitti. Nixon drove even more slowly, talking and laughing freely. They seemed to be enjoying their afternoon out of the office.

Sakau originated as a drink when people saw rats become dizzy and sluggish after nibbling on the root. The effects of *sakau* are often described as calming or numbing. Bill Raynor, who had published several papers about the plant, wrote in the book *Dangerous Harvest: Drug Plants and the Transformation of Indigenous Landscapes*, "The psychoactive effects of kava [*sakau*] are, in general, mildly narcotic, soporific, diuretic, and muscle relaxing." They are not, as members of Captain Cook's crew described, opium-like or hallucinogenic.

Before hibiscus bark was used to squeeze *sakau*, young virgin women squeezed it in their long hair. High chiefs would kneel at their feet and drink it as it cascaded down their smooth brown legs. When the Catholic missionaries first came to the nearby island of Kosrae, they forbade the drinking of *sakau*. It is widely believed that *sakau* root was smuggled to Pohnpei in a woman's vagina—enough, at least, to start a new plant.*

I asked Marciano why the drink had survived in Pohnpei. Hadn't the missionaries tried to forbid it there as well? "Pohnpeians are more stubborn," he said. But in general, the reason Pohnpei had retained its early traditions better than neighboring islands was because more of its people survived smallpox.

Kosrae's population was almost wiped out when whalers from New England first brought the deadly disease. With only about two hundred survivors, the Kosraean culture had broken down. But for some reason in Pohnpei about two thousand

---

* Because of this, Marciano said, "the *sakau* in Pohnpei has an odor, but the *sakau* in Kosrae does not."

people survived (around 20 percent of the population), enough to carry on customs and resist the advances of the Spanish missionaries who arrived soon after the devastation.

Marciano said that most Pohnpeians were Catholic but practiced a kind of hybrid spirituality, incorporating native beliefs with Christian ones. Bill Raynor had originally come to Pohnpei as a Jesuit volunteer and soon learned that the indigenous faith had never gone away, just entangled itself about the church like a tenacious vine. He told me later in the trip, "Anyone who's a straight, Western-style Christian should stay off the island." Such an integrated culture left room for eels (which could easily be assimilated as the snake figure in the Garden of Eden).

At last we arrived in the municipality of Kitti and the pool of eels at Pwodoi.

Beside the street, above the creek, was a small stand where boys sold betel nut. The oval palm seed, locally known as *pwuh*, was chewed with lime (usually crushed coral) wrapped in a leaf of a native pepper plant, called *kapwohi* (*Piper betel*), and produced a mild stimulant effect (as well as a bright orange-red fluid).*

Nixon and Marciano bought some betel nut, and the young vendors led us down to the creek. We stood at the edge of a large pool where the water passed under the street through a culvert. One boy had a can of mackerel and punctured the top with his knife, allowing fishy juice to leak into the water. A few

---

* Red mouths, spit, and teeth produced by chewing betel nut would be a common sight to anyone traveling in parts of Southeast Asia.

large shapes began to emerge from under the broad-leafed foliage overhanging the brook, and as they stirred across the light sandy bottom into a ray of sun, I saw for the first time the beautiful golden color and brown mottling of the tropical freshwater eel, *Anguilla marmorata*.[*]

One of the boys stepped into the creek and put his hands beneath the belly of a particularly large eel. Even Nixon and Marciano cringed slightly, as the eel was longer than the boy was tall. But the big fish allowed the boy to caress its body, even lift most of its body out of the water. Pohnpeians maintain that only those of the Lasialap clan can safely hold a large eel; anyone else who tried would be bitten. When I first heard this I was skeptical, as I had touched, petted, and even dived with large eels in New Zealand and never once been bitten. What was remarkable was how the boy was able to hold the eel out of water for so long with so little protest from the fish. I tried this later and was un-

---

[*] *Anguilla marmorata* is native from South Africa through Indonesia and New Guinea all the way to French Polynesia and north to Taiwan, Hong Kong, and southern Japan. It is the most widely distributed freshwater eel species in the world. As with other species of freshwater eels, no one has witnessed the tropical eel spawning in the wild. Any estimates of spawning locations are based on the capture of *A. marmorata* larvae. There are thought to be at least five different populations of this species, each with its own spawning area—at least one in the Indian Ocean, two in the western North Pacific, and two in the western South Pacific. Recent genetic and morphological data indicate that the Micronesian eels of this species are a separate population, distinct from all others, but its small larvae have yet to be collected.

able to do so. There was something special about the relationship between these people and the eel.*

～

The following morning Bill arranged for me to meet Adelino Lorens, who besides being the minister of agriculture was a high chief of the Eel Clan (with the customary high title, *soulik en dianso*) and a deacon in the Catholic Church.

Bill had worked closely with Adelino on the identification and protection of Pohnpei's indigenous plants and on the Island Food Community Project, which promoted the benefits of eating native fruits and vegetables. Bill and Adelino had identified more than forty native varieties of banana, some of which were endangered, and had posters produced that helped people identify the ones that were the most healthful to eat.

Adelino's office was in the Ag Station just beyond the Visitors' Bureau. He greeted me there, a handsome, soft-spoken man in his late fifties with short-cropped white hair and a gracious smile. We sat down in chairs at a long table as a brief torrential rain drummed loudly on the roof.

"I've been in touch with Sahngoro, the high chief of U and the Eel Clan," Adelino said. "But Sahngoro deferred to Resio Moses, the *tauk,* or third chief, of the Lasialap. Resio is a senator in the Micronesian Congress, a very busy man. He said he

---

* Although people of the Lasialap clan were from U, some Lasialap lived outside of that municipality in other parts of the island, such as in Kitti.

won't be able to meet with you during your stay, but suggested you meet with an elderly woman named Ester Alex."

There was a long pause. A warm breeze blew through the room where we sat. Steam rose from the wetness on the ground under a suddenly hot sun.

"Is it that they are busy," I asked Adelino directly, "or do they just not want to meet with me?"

Adelino turned his head to one side, considering my question.

"I think maybe some individuals believe that another person may know the stories better, so they defer to them," Adelino said gently. "But also, you know, in Pohnpei, people are reluctant to share traditional knowledge. You can pass all that information to any member of your family when you are ready—and you are ready when you are near to dying."

Adelino told me Pohnpeians believe that if they tell a complete story from beginning to end, they will die. Knowledge is a kind of energy that flows out of the body, and if you share it all, you become weak. That is why people usually share all their knowledge only when they know they are about to die.

"But you should have no trouble getting pieces of the stories about eels," Adelino said on a more encouraging note. "And then maybe you can put those pieces together."

I realized my approach had been somewhat presumptuous. Walking into a village and asking someone to share a traditional story was like walking into a workshop and asking a master carpenter to share all his or her trade secrets. But that did not diminish my enthusiasm or intent.

I asked Adelino, in his role as a deacon, how the arrival of Christianity in Pohnpei might have altered or weakened the indigenous faith.

"The church uses a lot of traditional knowledge," Adelino said. "They have to in order to break through to the people. For example, we use *sakau* two times a year in the Catholic Church: before Christmas and after Easter. But there is more. The beliefs exist together inside and outside of the Church. The Lasialap, though Christian, still consider the eel to be a human ancestor."

The last person to practice pure Pohnpeian religion died in the 1950s, but according to Bill and Adelino, the people still make the old religion fit. Plant magic is still practiced. One of the CSP employees, Valentin, is a deacon in the church in Madolenihmw but also an expert in the use of plants, specifically children's medicine. In the Pohnpeian language every plant has both a common name and another that could be described as its spiritual name. The spiritual name is known to only a few people and gives them influence over the healing powers of that plant.

～

That afternoon, Serlene, the office manager at the Nature Conservancy, took me in her car to Palikir to see Dr. Rufino Mauricio, the island archeologist and historian. Serlene was lithe and pretty with olive-brown skin like a Gauguin woman. As she drove, she talked about her aunt's farm and all the animals they had, and the delicious smoked bacon she made from their big, healthy pigs.

"I like the taste of dog most of all," she said, raising and then lowering her eyebrows in succession—the Pohnpeian manner of indicating the affirmative (as opposed to nodding the head). "More than pigs," she added gleefully. "They are *really* good. We raise dogs as pets, but we usually don't eat our own dogs. We eat our neighbors' dogs." She laughed again. "If a dog ever bites someone or hurts someone, we eat it."

"Would you ever eat eel?" I asked.

"Never!" she said, shocked at the mere suggestion, laughing at me.

Palikir, the capital of the Federated States of Micronesia, is not a city at all or even a town; it's a campus of government buildings ringed by cement columns molded and painted to look like the basalt logs of Nan Madol, one of the only outstanding stone ruins in the Pacific.

Mauricio met us in the Department of Archives and History and took us upstairs into a small library, where we sat across a table from him. He was gracious with his time and did not seem hurried.

"The eel came to Pohnpei in the seed of a plant that ended up in the ocean, in the reefs off the island, and it traveled in the water up the river, in through the mountains," Mauricio began, a bit ambiguously. "As it traveled, it charted the lineages of the Lasialap clan, giving birth to different subclans."

Mauricio said that it is typical for the history of a clan to be spun around the life history of an animal. He did not want to speculate as to how this fish had come to be important to the people, but somewhere along the way members of this clan became quite close to the freshwater eels. "And I think they actually take care of them," Mauricio said, "if I'm not mistaken. They consider the eel, you could say, their totem."

Though Pohnpei had no book with traditional stories about eels, Mauricio assured me that such tales existed. Even for him these stories were difficult to collect—not only because they were closely guarded, but also because they were disappearing. The younger generation was distracted by media other than oral storytelling—television, radio, and the Internet.

"It's a challenge our office is concerned about," Mauricio said. "Like the rest of Micronesia now, we on Pohnpei are experiencing rapid change—thought-wise and in attitude." One bright spot was the advent of CSP and the young Pohnpeian staff who cared about preserving the island's wilderness as well as the traditional knowledge of native plants and animals.

Mauricio felt that Pohnpeians had always been careful stewards of the land. He believed that one reason they built Nan Madol on the reef and not on the island proper was because the land to them was sacred and not to be disturbed. That was also the reason he gave for the people eating sea fish but not freshwater eels. "I think for the most part in Pohnpei we don't bother the eels too much," he said. "We just let them live. They always say, for a stream to flow year-round it must have eels."

Before Serlene and I left, Mauricio offered another tidbit of an eel tale.

"I know in one of the stories," he said, "a large eel had a place in a tidal channel, and when villagers went out in their canoes to go fishing, she would appear and ask, 'How many people are in this canoe?' So they would say how many, and she would ask them to drop one as a kind of fee so they could pass. This kind of story is from Madolenihmw."

"Do you know any other stories like that?" I asked.

"One other thing that I've seen," Mauricio said. "Sometimes when it rains, the heavy rains, you'll be walking along the trails and you'll run into small freshwater eels. So people would believe that maybe these guys fell down from the sky. I think maybe the water level of the river rises up and the eels get stranded. I've seen them far from the river, in the roots of trees.

"You know," he continued, "some years ago there was one young woman, their house was right on a stream. People knew that she had cancer and was dying. People would spread rumors like, 'In the morning we found a freshwater eel with the girl, dead.' I don't know what the point of that is. But every once in a while you have a story like this."

⌒

The next day, I stopped by the offices of the CSP to say hi to Nixon and Marciano. They were out, but a staff member named Leinson introduced himself, and volunteered to take me to see one of the island's tallest waterfalls, called Sahwartik, on the Lehn Mesi River. Then we could have *sakau* in his village, Enpein, in Kitti and try to find people willing to tell us eel stories.

"I think it is interesting, what you are trying to do," he said. "Maybe I can help you."

Sahwartik Falls was on the west side of the island near Leinson's village. To get there we had to take a small dirt track that climbed the mountain from the coast and the main ring road. The road got smaller and smaller the higher we went, and the jungle denser. It felt worlds away, even from little Kolonia.

We parked the car and hiked down a steep hill, holding on to branches of trees and vines as we went. The Lehn Mesi River, Leinson said, had been carved out of the ground by the body of a giant eel. We could hear the currents tumbling over stones at the bottom of the valley.

Turning leaves with his feet and exposing rich dark soil, Leinson showed me how, imperceptibly, we were walking in a "farm." Crops—*sakau*, yam, and taro—were embedded in small clearings in the forest. What the untrained eye would never recognize as cultivation is what Bill had described to me as agroforestry.

Further up the trail we could hear the crashing sound of Sahwartik Falls, and as we drew closer we could feel the cool mist breaking the humid forest air and then see the tall stream of water falling. Leinson took off his shirt and jumped into the deep pool at the base of the hundred-foot falls. I followed, and after a swim we sat on the rocks near the pool in the cool spray and watched a black tern-like seabird called a noddy circle the pool in broad turns. It felt strange to see an oceangoing bird on a freshwater stream, but the sea, I was reminded, was never far away.

It was getting on toward four o'clock and Leinson was itching to get to the *sakau* bar. We left the falls and headed back up the trail to the car.

As we walked, Leinson reiterated the importance of *sakau* in the culture of Pohnpei.

"If there is a tragedy—for instance, you hit someone's child on the road with your car and kill them—you meet with the family and drink *sakau*. It is the only way to get forgiveness. If you meet someone you want to marry, you ask the entire family permission over *sakau*. If you're sharing an important story, it is done while drinking *sakau*."

Along the road back down to the sea we stopped at a *sakau* bar in Enpein. Being from that village, Leinson knew all the men and women gathered under the shelter of a thatched roof near the *peitehl*. He introduced me to the village chief, Herbert Mikel, a former Micronesian senator. Leinson told Mikel why I'd come to Pohnpei and that I'd come from the States. Mikel asked where; I said Connecticut. Mikel nodded. He rubbed his chin, taking it all in. He told Leinson, in Pohnpeian, to tell me that he had a brother who taught at Eastern Connecticut State University. All the women were laughing. Mikel spoke perfect English but wanted Leinson to get used to translating for me— if he was to take me around to meet the old people who would tell us stories about eels, he would have to sharpen his skills. Mikel gave Leinson a tidbit of a story, about the river called Lipwentiak.

"The eel went up the river, and saw a man-eel with a belt full of birds, and she got scared and turned around and carved a circular hole in the river."

That was it. Everyone was laughing and passing around the coconut shell full of *sakau*. Two men behind the *peitehl* were stripping hibiscus bark. Two more men at the *peitehl* were pounding *sakau*. Mikel offered me a drink. The *sakau* in Enpein was extra gelatinous, Leinson explained, because they used more hibiscus bark to render the drink. People seemed to be getting pretty stoned, the women and the men laughing, laughing. I had drunk only enough to get a mild effect. I had not yet drunk the amount required to make my forehead and spine go numb, as was described in the literature.

Some corner of my mind heard a parrot cackling. When I turned around there was a beautiful young girl standing there, a deep mauve bird with a bright yellow beak perched on her hand. It spoke in her ear. If you were an avid birder and were dropped unaware into this scene, you would know this was the island of Pohnpei, because the Pohnpei lorikeet lives here and nowhere else. There was something magical in that, something so specific to place, very much like the stories we would hear in the coming days.

Eventually Leinson and I said goodbye to the chief. I had promised Bill that I would join him for dinner and *sakau* that evening in Kolonia. Sunday, in the early afternoon, I would meet Leinson back in Enpein.

"I will ask some people about stories," he said.

"How will I find you?" I asked, as he didn't have a computer or cell phone, or even a land line. He waved his hands in the air. "It's a small village. Everyone knows everyone. Just drive up and ask for Leinson Neth."

He walked into the forest, and I drove back to Kolonia.

I was late getting back into town. I stopped to get Bill at the Nature Conservancy offices. He said that if we were going to the *sakau* bar we wouldn't have time to sit down for dinner, so we picked up some fried chicken on the way. There was some urgency getting to the *sakau* bar, he said, because the first squeeze was the most potent. As the evening wore on and the root was repeatedly mashed and squeezed, and water added slowly, the drink became thinner and weaker.

We met up with a friend of Bill's named Tony, someone from Bill's former life when he first arrived on the island as a Jesuit volunteer in the 1970s. Even though they lived on the same small island and had once been best friends, they hadn't seen each other in years. We sat at a long table with other local people, and as the sun went down, cups of *sakau* were passed and refilled and passed. Stories rolled out, volleyed back and forth from either side of the long table.

Bill had been a wayward youth looking to get as far away from home as possible. From the time he was two years old, he said, he knew he was going to live somewhere else. When Bill first moved to Pohnpei from California and worked as a teacher at a local high school in Madolenihmw, the road around the island had not yet been built. To get from Kolonia to Madolenihmw you had to take a boat. Telephone numbers were only three digits. He drank a lot, slept with a lot of women, and got into trouble with his boss, the late Father Costigan. But as months and then years went by, he fell in love with the people and the place, especially the agricultural way of life. He learned that although the people lived by the sea, they were not seafarers; they ate sea fish, but primarily they were people of the land. A self-identified "aggy" who'd worked back home on artichoke and pumpkin farms, Bill became fascinated with the Pohnpeians' gift for growing things and their knowledge of native medicinal plants. "Here was this place where the most celebrated person was the one who could grow the biggest yam," Bill said. "That was for me."

He got his master's degree at the University of Hawaii, studying open-pollinated plants native to Pohnpei. In the course of his research he walked a good part of the island, from the lowlands to the cloud forest at over two thousand feet. He slept under the trees, contracted rare diseases (including elephantiasis), went to every funeral, counted every yam, measured every tree, and identified every plant. He learned the language, married a local girl (the daughter of a high-ranking chief), and never went back to California except to visit his parents or for the Nature Conservancy.

"He's the guy on the island who knows the names of all the breadfruit and banana varieties, the endemic palms and cinnamons," Tony said of his friend Bill.

"That's how I got my title, *sou madau*," Bill said proudly, "master of thinking and planning. By growing a big yam."

Tony passed around a bottle of cheap whiskey to double the effect of the *sakau*. As the coconut shell was passed, the discourse flowed like a river—sometimes trapped in a tangled eddy, sometimes ripping in flood, but never returning to where it had been.

"The virtue of the coconut shell as a cup for *sakau*," Bill said, touching its rounded underside, "is that you can't put it down until it's all been drunk."

My mind turned inward and I could no longer talk. I thought about the whalers, missionaries, pirates, and tattooed Irishmen who had come here to this island—the Spanish, the Germans, the Japanese. Some thousand years before, the Pohnpeians had constructed the mysterious stone palace of Nan Madol. Those people (their genes at least) were still here, shedding waves of settlement like dead skin.

✦

I was a little slow the next morning, everything still moving at quarter speed. I sat on the balcony outside my room, took a deep breath of the salty humid air, and sketched a few items I had collected—a seed of the *oah* palm, the leaf of *Barringtonia asiatica*, a native banana with a rusty orange peel called *karat*. I ate a greasy Pohnpeian donut and drank a few cups of coffee. I was still feeling languid. I had forgotten home. I myself felt endemic.

Bill was in pretty bad shape from the drinking the night before. He had to interview a few people for a job opening and was not looking forward to it. He said he had some good news on the eel front, though. Adelino was going to take me back to his village, Awak, that afternoon, and his son Allen would accompany me to see Ester Alex, the elderly woman who'd been mentioned to me as having an eel story. I had yet to hear a story that was more than a fragment.

I had finally come to terms with the fact that daytime was dead for storytelling in Pohnpei. So I loitered near the markets by the water, watching the fish come in—yellowfin tuna, in whose bulbous bodies and globe-like eyes you could read the sea. The fishmongers were cleaning the tuna and throwing bits of fish heads and guts into a drain in the sidewalk. As I walked by I heard a mysterious splashing and slashing. I could not help but look down into the darkness below the steel grate. There were eels, big ones, in a small bit of moving water in the street gutter.

I continued up and down the streets of Kolonia, admiring handicrafts and other curiosities, and before I knew it, it was time to go see Adelino at the Ag Station.

We drove, with one of Adelino's employees at the wheel, the twenty minutes from Kolonia to Awak. The way around the east side of the island was more exposed to the sea than the drive around the west side to Kitti.

Awak was a sea-blown village, with broad views of mangrove islets. We pulled into the sandy lot of a weathered blue and white Spanish-style church—the church of which Adelino was deacon. I insisted that the driver take some money for gasoline, which was quite dear on the island. Adelino acknowledged my offering as very generous.

A beautiful creek tumbled down the mountain behind the church, breaking into two branches that flowed around it, embracing it on its own little island. We stood on a bridge on one side of the church looking into the currents. Adelino pointed to some cement pilings on the bank and a monster eel lying straight and still between them.

"When I was a child," Adelino said, "the old church was here, on these pilings. During services, we could hear the sound of the river under the wooden floorboards. The eels were never far away."

The existing church, surrounded by tall palm trees, gleamed white against a menacing purple-black storm cloud. The ocean behind us was blue and choppy.

Leaving my bags in the church parking lot, I followed Adelino to the home of an old man who lived beside the river. His name was Manuel Amor. The man pointed out that the river here, the Kepin Awak, was tidal, and that the big eels could come and go as they pleased. He said when the eels got really big they went to the ocean and never came back—he believed that they were eaten by big sharks. He had a bowl of dishwater with a bit of fish meat suspended in it, and as he poured it into the river, big eels with wide heads, long nostrils, and strong, muscular bodies appeared from the sides and bottom of the pool. Some village children materialized from behind the house and stood at the edge of the pool. I had an underwater housing for my camera, and held it just under the water where the eels gathered, to take a few pictures. The eels were curious and nosed the camera housing and my hands. I thought it would be really embarrassing, not to mention painful, if one of the eels grabbed hold of my hand, so I pulled the camera out and stepped back from the pool.

Adelino next took me to his home. He said he was very busy preparing for services on Sunday, and in addition, the next day he was entertaining a visiting archbishop. But he was gracious nonetheless. We sat down on the front steps of his home, half hidden in the jungle. He said that his son Allen would be arriving soon and could walk me from the church down the road to visit Ester Alex.

Adelino leaned forward and asked me how I had come to find out about Pohnpei and the Lasialap people. I told him I'd been in coastal Maine doing research on the glass eel fishery when I met a Taiwanese eel dealer named Jonathan Yang. Yang first told me about the island, and shared a story about his friend Mr. Chen, who had died mysteriously after trying to export big eels.

Adelino nodded, as if the story meant something to him.

At that point Adelino's son Allen arrived, a handsome, wiry young man. Adelino excused himself and Allen and I began our walk along the road to the home of Ester Alex.

In the waning sunlight sea birds called vigorously from the jungle on the west side and the deep mangrove thickets on the eastern, ocean side of the street. The homes of the people, nestled in the jungle, looked ragtag and simple. Most Pohnpeian homes are open on the sides, with a wall that may go as high as the knee—the people sleep in the open, surrounded by nature. Within, the homes are clean, pots stacked neatly, cement or dirt floor swept. I walked with Allen for about a mile, and by the time we got to the home of Ester Alex, it was nearly dark.

Her house was very small, with a corrugated metal roof held up by tree fern posts on a poured concrete floor. Erlinso, Ester's son, put out a seat for his elderly mother on their front porch, then sat next to her, translating as she spoke. I was excited to be hearing my first eel story from a Lasialap. The soft chortling sounds of caged lorikeets, the voices of children, and the night sounds gave texture to the air. Erlinso, a schoolteacher in Awak who taught English, spoke softly and slowly, with many pauses. The pauses were partly to let his mother complete her sentences, but also, it seemed, to let the breeze and the night sounds flow freely in between the phrases.

"You know, there was a story about eel." Erlinso pronounced *eel* with an *h*, "heel." "A long, long time ago, there was a couple in Kitti. One day the wife went out fishing with a fishing net. So when she dipped the net in the water and lifted it up, there was a small rock, or stone, in the net. She took the rock and threw it away, and she moved away from it, maybe ten or twelve feet away. And then she put the net in again. When she lifted it up, the same stone was in there! And then she picked up the stone again, and threw it away, and she moved to another place. And the third time she put the net in the water she lifted up the same rock." His voice expressed the amazement and exasperation the fisherwoman must have felt when she pulled up the rock a third time. "And then she put the rock in her basket and went home. When she got to her house she showed the stone to her husband. And the husband advised her to put the stone in a small well— they used to drink from that well. And then one day, they found that the stone had cracked open, and a small fish came out—an eel.

"They used to feed the fish in the well," Erlinso continued, "and when it got bigger, they decided to kill that eel. And so, un-fortunately, the fish heard about their plan. They planned to kill the eel, but the eel heard them . . . she was a ghost. So she moved from there, from the well, into the forest and finally landed in Madolenihmw, and she gave birth to a woman, and that woman was Lasialap. And so we are descended from that woman. Our clan is known as Lasialap." He paused, and it seemed as though the story was over. Then his mother spoke again.

"And then the eel went out to Kosrae," he said, translating. "Have you been in Kosrae? The eel left Pohnpei and then went to Kosrae, and then, in Kosrae, the eel gave birth to another woman. That woman came back to Pohnpei. The *nahnmwarki* of U, the uncle of Adelino Lorens, he is descended from that woman that came from Kosrae. Then, after that, the eel moved from Kosrae to Pingelap and then from Pingelap to Yap, and then from Yap she came back to Pohnpei and died in Nett.* There is a mountain in Nett. If you look at that mountain, it looks like an eel.

"So, we the Lasialap, we really, you know, respect that eel. But you know one bad thing—some people, from Kosrae, they came out here and killed some eel to eat. If we didn't make a rule, then very soon the freshwater eel will disappear and all the waters will dry up." Ester stopped speaking and stared at the floor, her sunken mouth closed.

---

* Pingelap is an atoll of Pohnpei.

"So, the freshwater eel is a human," Erlinso continued on his own. He wiped his face with a rag and lifted his eyes to look at me. "You know, they used to whistle just like you."*

"The eels whistled?" I asked, wanting to make sure I'd heard correctly.

"Yes. The people, they get mixed up, they thought people were doing that, but it was the eel. You know, when the *nahnmwarki* of U dies you will see the freshwater eel walking on the road. Just the day before the *nahnmwarki* dies, usually we see the eels dancing on the road, and that's a sign that something will happen, especially to high-ranking people. We the Lasialap, we don't play with the eel. We really honor that fish. When the sun is too hot and the small wells are drying up, sometimes we take the eels and bring them to the ocean so they don't die.

"Well," Erlinso said suddenly, putting his hands on his knees and motioning for us to get up. "Thank you for coming." He shook my hand.

Allen and I walked back to the church and his father's home on the dark road. Everyone in the village, it seemed, was out walking on the road: boys and girls, young men and old women, dogs and chickens. The street is a village, a social landscape.

On the way back, Allen spoke openly with me. He told me that his grandfather had been the *nahnmwarki* of U, and when he

----

* In his *Dictionary of Celtic Mythology*, James MacKillop writes, "In the west of Ireland whistling eels were thought to foretell famine. Yet at other times eels might be benevolent, such as those thought to be the guardian spirits of wells and magic springs."

died, they went up into the mountains and dug a big *sakau* plant for the ceremony. "I was just a boy," Allen said, "but I was old enough to go with them. When they brought the *sakau* back to the village, here to Awak, they prepared to chop the big root in half. As the man raised his machete to strike the root, an eel appeared. I saw it, standing up on the tip of its tail, in the middle of the stump. The people carried it down to the river and let it go."

Adelino had told me that he was spending the evening with his family at a retreat by the ocean. He said they got together a few times a year, ate, drank *sakau*, and camped by the ocean, but he did not ask me if I wanted to come. As Allen and I continued on the road to the church, I was wondering when he was going to say goodbye. But we continued talking and had gone beyond their home, and soon turned onto a sinuous path made of piled coral that wound through the flooded mangroves. I was surprised when Allen mentioned that we were going to meet up with his family, and I wondered when his father had had a change of heart and decided to have me there.

The mangroves were dense and it would have been hard to see the path were it not made of bleached white coral. There are no sand beaches on the island of Pohnpei (part of the reason it has never become a tourist destination), just on the outer atolls; the edges of the island are mostly mangrove forests that flood at high tide, so if you want to be near the ocean, you generally have to build a platform of coral or stone.

We came to the other side of the mangrove thicket and the campsite where Allen's family was spending the night. A ceremonial open-air structure called a *nahs* was perched near the water, and Adelino and his extended family were seated under its roof around a *peitehl*, pounding and drinking *sakau*. Adelino welcomed me and introduced me to what he described as his entire extended family. He asked me to sit next to him. All the men were shirtless, and I followed suit. The two young men at the peitehl were preparing the hibiscus bark to squeeze the *sakau*. One laid out the long strips of hibiscus bark, folded at one end. He evened out the fibers of the slimy bark on the lava stone and loaded it with a mound of crushed root, then held the bark at both ends and twisted, squeezing the root wrapped inside. At the end of the squeeze the man was straining his muscles as the other held out the coconut shell to capture the fluid. The coconut shell was passed to Adelino, and Adelino passed it to me. He said, "This is the fourth cup. We are honored to have you here. Please drink."

Beneath the *nahs* was spread a feast, but it was all covered and it didn't appear that anyone was eating yet. The cup was filled and passed and filled and passed. As we drank more, conversation slowed. I wondered what had encouraged Adelino to ultimately invite me to this gathering. It could have been that he finally understood my passion for the eel—or it could have been that he wanted to somehow repay me for giving his driver $20 for gas.

"We knew this man from Kosrae," Adelino told me, "who came to Pohnpei and married a Lasialap. One day, when his wife was away, he could not resist—he killed a big eel and ate it. The next day he felt sick, and a month later he died."

Adelino did not speak further, just stared. The night was silent now. People had drunk enough that they stopped talking.

Later in the trip I heard the same story about the Kosraean who ate the eel, but from a different source, a woman named Shelly who was born on the Pohnpeian atoll of Mwoakilloa. She said that her husband's auntie had married a Kosraean. He lived in U because his wife was Lasialap. He saw the eels in the stream near their home and always talked about eating them.

"His wife wouldn't let him," Shelly said. "She was really against it, she didn't want him to eat them. But one day he went behind her back and caught one eel and cooked it and ate it, but didn't tell his wife. A few days later, he started to get sick, and ended up in the hospital. His sickness only got worse, so they took him home to Kosrae, where he died. They said he had spots on his skin like a rash that resembled eel skin."

Serlene, at the Nature Conservancy offices, followed Shelly's story with her own. One day when her husband was a kid he was out in a stream and decided he'd try to kill eels for fun. He speared one, then another, and he was about to spear a third when he fainted. They took him to the hospital, where he was very sick for days. They said his vomit smelled like eel.

I did not want to overstay at Adelino's family gathering. I got up and took a walk by a kind of man-made enclosure, a pen of water circled with coral. There were a few big sea turtles in the water, and children ran around the periphery, chasing the turtles, which swam gracefully in the moonlight.

Bill had spent the morning readying soil for planting and was covered in dirt and sweat. He'd inserted plant cuttings at intervals in the dark mounds of soil. The cuttings were wilted, and I wondered out loud how they would recover.

"We don't water anything," Bill explained, grabbing a hoe from the soil and leaning it against a tree. "Any time of year, it is sure to rain soon." Rain it did, and Bill's cuttings rose from limpness toward the sky.

The night before I had walked to the Village Hotel in Awak, clambered to my room—a freestanding hut in the jungle—in the dark, and woke to one of the most beautiful ocean views I had ever beheld. From the ridge was a stunning view of Sokehs Rock, far in the distance, and in the foreground, in the tops of the palm trees, cardinal honeyeaters flittered in and out of the fronds.

Bill's house was in the municipality of Madolenihmw, at the opposite end of the island from Kolonia. He had invited me to stay the night, go to church with his family the next morning, and then visit Nan Madol.

We walked down a steep path below Bill's house. It was jungle, but not dense, with taro, tomatoes, *sakau*, eggplant, yams, different varieties of banana, breadfruit, and avocado. We stayed out for part of the day through several heavy showers. As the sun was setting and a delicious sea breeze started to blow, Bill walked me back to his house.

Bill and his wife, Beli, lived in an open communal-style

home of cement blocks covered in plaster and painted salmon pink. Beli is a dressmaker, employing two seamstresses. She is a direct descendant of the *nahnmwarki* of the Breadfruit Clan. Besides the house they also have a large open *nahs* for big family gatherings.

In the home and the *nahs*, children, men, and women had gathered just before dark. They ate blue parrotfish, taro, chicken, and rice out of bowls with their hands, and staked out territory for sleeping.

"If you're not a social person," Bill said, "you won't survive in Pohnpei."

A few young men appeared with a *sakau* root and began washing, cutting, and pounding it on the *peitehl*. Simultaneously, children and adults began watching a Japanese horror movie on a television hooked up to some remote power source. You could not escape *sakau* if you wanted to.

———

The next morning we walked to the church, across the street from the school, now closed, where Bill had had his first job on the island. Like the church in Awak, this was on the ocean, and the doors and windows were wide open, allowing the fresh morning breeze to flow through. I recognized the man leading the service in front of a large congregation of adults and children in colorful blues and greens: Valentin, an employee of CSP whom Bill had said was knowledgeable about plant magic. Bill had explained that there were three main divisions of social structure, church, government, and the traditional system, and "if you are high-titled in all three, you've got it made." Valentin was one such man.

After the service we headed for Nan Madol, the stone ruins sometimes called the Venice of the Pacific (because canals of sea water once coursed between the buildings). There was no sign indicating the turnoff from the main road. Bill drove down a weedy track to a small kind of house that belonged to the family who claimed to own the land and the reef beneath the ruins. A man collected a few dollars as an entrance fee, and Bill asked him in Pohnpeian about eels.

"That guy," Bill said, "he says he has a story of a big eel that lives in the mangroves. But that's all he will say."

Nan Madol was the political and religious seat of Pohnpei under the Saudeleur Dynasty (ruled by one man, the Saudeleur) until at least the sixteenth century. Covering 150 acres, Nan Madol consists of ninety-three man-made coral islets. The area was occupied as early as 200 B.C. but the structures of giant basalt crystals that make it one of the most magnificent architectural feats in Oceania were probably not built until the twelfth century.

It is said that priests fed dogs and turtles to a giant moray eel that lived in the coral footings. The eel, named Nan Samwol, was considered a guardian spirit of Nan Madol.*

---

* Bill suggested to me that as island culture shifted from seafaring to terrestrial life, the freshwater eel took on greater importance than the saltwater variety.

"At a time determined by divination and the change in agri-cultural seasons, the priests performed an extended ceremony of homage, supplication, and atonement called Pwung en Sapw. The ceremony culminated in the offering of a turtle as tribute to Nahn Samwohl, the great saltwater eel that dwelled in a shallow pool on the islet of Idehd in Nan Madol. Nahn Samwohl's acceptance of the offering indicated that Nahnisohnsapw [the principal god of the *saudeleur*] was pleased with the conduct of human affairs on Pohnpei." So writes David L. Hanlon in *Upon a Stone Altar: A History of the Island of Pohnpei to 1890*.

Nan Madol

It is believed that the pentagonal and hexagonal basalt crystals (some twelve to fifteen feet long) that make up Nan Madol were quarried from lava outcrops at distant parts of the island. A cornerstone on one part of the ruin weighs more than fifty tons, the average being well over ten, so modern people have naturally wondered how the ancients got them there. Some say they moved them on bamboo rafts, but bamboo, according to Bill, is not native to the island. The islanders say that a magician flew the stones across from Sokehs Rock, where much of the basalt was likely quarried. Archeologists have tried to determine what kind of rafts they may have floated the stones on but have come up with no feasible method—certainly none that's been successfully tested.

"The best explanation I've heard," Bill said, stopping on the path in the mangroves, "is that there was a lot more magic here back then."

Bill had kept up a running commentary about Nan Madol history as we traipsed along the path through the jungle, but once among the ruins, now overgrown with palm trees, vines, and mangroves, we walked in silence.

It was a warm day, and it was refreshing to wade around the ruins in the shallow lime-colored water. The palace buildings were extraordinary, the texture and patina on the basalt "logs" making them seem almost like a dark steel sculpture. For a while I lost track of Bill, and I was alone among the breadfruit trees and palms and the bright red cardinal honeyeaters.

Just after noon on that clear bright Sunday, I left Bill's home and drove around the south side of the island to Kitti to meet with Leinson Neth. Among the CSP employees, he was the only one who seemed genuinely keen and determined to help me. I think that was partially due to his own curiosity about eels and their deep and old relationship with his people.

I drove into Enpein as Leinson had instructed and asked a group of boys in the street where I might find him. Within minutes Leinson came walking down the street toward me. He had a plan: we were going to see his uncle.

We drove to the church grounds, where people were getting out of midday services. At the deacon's house we found Leinson's uncle, Lorenso Gilmete. He was seated on the ground, in the shade of the front porch, legs folded, picking at calluses on his feet and grumbling in a low, gravelly voice. He appeared to be drunk, periodically licking his lips. This did not look promising.

Leinson, good-natured and kind, motioned for me to sit near his uncle. He mentioned in English and then Pohnpeian why I had come to Pohnpei.

"Hmmm," his uncle grumbled, then laughed. He said a few more words in Pohnpeian, and laughed again. I got the feeling he was mocking me. He said something to his wife, Leinson's aunt, standing nearby, and she laughed, too. Leinson kept a stone face. I asked him what his uncle was saying. Leinson was probably too polite to say exactly.

Lorenso motioned for a young boy to bring some food. The boy made a plate for me of boiled yam. It was white and semi-sweet, mashed with coconut milk. The old man laughed once more and asked me if, in my travels, people had been reluctant to share their stories. I told him yes, that some had. I respected that he might not want to share, though inwardly I would be deeply disappointed.

The old man leaned back against a wooden railing and asked if we could talk about eels the next morning. Leinson looked at me and shrugged. I put on my most disheartened look. Leinson threw a pebble into the dirt. The uncle changed his tone.

"Do you like music?" he asked me in English.

"Yes," I said. I thought he was going to play his radio, but instead he started rocking forward and back on his haunches and broke out in song. In a lilting cadence, an earthy bass, he sang in Pohnpeian for about a minute and then stopped.

When his uncle had finished singing, Leinson smiled. He said the song was about *kemisik*, the eel. "It is a reminder of how eels first came to Pohnpei," he said.

The song was the prelude to a story; the song and the story fit together like a puzzle, one part being insufficient without the other. "The song refers to what is to come," Leinson said, adding that it was very hard for him to understand some words because the song was in local dialect.

Leinson's uncle followed the song with a few words to his nephew.

"He wants us to give him a ride to the *sakau* bar," Leinson said. "And then we will take him to his house."

"Did he say if he was willing to share a story?"

"No, but I think we should do as he says."

We drove Leinson's uncle and aunt to the *sakau* bar. He said to come back in a few hours, right before dark. While he was drinking, Leinson and I would go up a small road into the forest to visit an old woman who, he'd heard, lived by a stream with a monster eel.

We left the car on the roadside and walked through the jungle up a narrow path that skirted a small stream. The pools in the stream were dark and shaded, the water cool and crystal clear. After about ten minutes we came to a clearing and a small thatched-roof home. An old woman, thin and frail, stepped out of the home to greet Leinson and me. She'd been expecting us.

She was very hospitable, offering us a seat and signaling to her granddaughter to make us tea. She herself sat on the ground, and so Leinson and I did, too. She told us that there was an eel in the small creek near her house that had become too big, and she feared that it might hurt her granddaughter. Her granddaughter, a young and beautiful girl, washed their clothes in the pool where it lived. The old woman said that eels were people.

"You see me, you see the eel," she said. "They are the same thing."

She told us that she had recently thought about trying to move the eel or even kill it. Her parents knew magic to move such an eel, but she didn't have the full knowledge—her parents, she said, had hidden part of the knowledge from her. She really loved the eel, but she was scared because she didn't have a full understanding of how to treat this creature, and feared one day she'd offend it and something bad would happen.

"My parents, they used to be able to talk to the eels," she said. "My mother could tell it to go away."

Leinson suggested that we go to the pool where they did their wash, to see the eel she was talking about.

When we got to the edge of the deep, narrow pool in the jungle, Leinson knelt on the bank and opened the top of a small can of mackerel, drizzling fish juice in the water. It was not long before a huge eel came out from under the roots of a tree. I had never felt very afraid of an eel before, but this one seemed erratic and unpredictable in its behavior. It took the food near the bank slowly and cautiously, then jerked its body backward into the dark depths. Another eel, about a third the size of the one we'd been watching, came up behind the big eel, vying for a flake of the white-fleshed mackerel. The giant eel turned suddenly and grabbed on to the smaller eel's side, twisting around, wrestling it, their bodies intertwining and frothing the water in the pool. Leinson stepped back and so did I.

"He's saying, 'This is my house,'" he said.

Once the eel had chased the smaller eel downstream to the tail of the pool and through the riffle into to the next pool, it returned to the bank, staring up at us expectantly like a dog, even pushing its nose up above the water's surface, showing its small eyes and horn-like nostrils. Though I had swum with big eels, I would not have gone into the pool with that eel; it was too strange.

Leinson and I walked back to the clearing to say goodbye to the old woman. We sat down with her momentarily. You could tell the eel made her nervous, like having an unpredictable neighbor.

She told us that many years ago her sister had an eel they were feeding and it got too big, so they decided to kill it. She thought the eel was becoming too dangerous for her children, and she was carrying another child. Her older brother came to kill it, and the eel chased him up the river. For days, no matter what pool of the stream he went to, even high up in the mountains, the eel was there. One day, he killed the eel. Soon after, her sister gave birth to a baby with closed eyes. When they brought the baby to the man in the village who had the knowledge of plant magic, he said it was because they'd killed the eel. Somehow he was able to open the eyes of the child. Leinson had heard that story before. The child in the story was the father of a CSP employee, Primo Abraham.

It was coming on evening when we began walking back to the car. The air had a peculiar stillness, a suspended weight, as we drove to the *sakau* bar where we'd left Leinson's uncle, Lorenso. He was sitting at a table with some other people, laughing and generally engaged with the group. When he saw us he asked if we could buy him a six-pack of beer and a few glass bottles full of *sakau* for the road. I did so.

Lorenso and his wife got up and came with us, and we rode in silence to their home. The old man instructed me to drive downhill from his home to the *nahs*. Like most *nahs* I had seen, it was a metal-roofed meeting area with tree fern columns and a U-shaped platform of cement, outfitted with a *peitehl* in the middle for pounding *sakau*. We parked the car and walked into the shelter of the structure. Lorenso made himself comfortable on the platform, seated as if on the stage of a theater, and set his six-pack of beer and bottles of *sakau* beside him. By this time it was dark, and Leinson turned on some lights around the *nahs*, one hanging precariously like a spotlight over his uncle. The jungle sounds grew louder and louder, as if someone was steadily raising the volume.

Lorenso asked Leinson to pass him a plastic cup, and poured some *sakau* for himself. The old man then told me to turn on my digital recorder, that he was about to begin his story.

He began, his eyes closed, with a chant. He sang the entire song that he had sung before, the song that preceded the story and was woven into it.

As Lorenso Gilmete told his story, I had no way of knowing what was being said because he spoke it in Pohnpeian. When I looked at Leinson, though, I could discern from his facial expressions that he was engaged with the story and found it meaningful. There were all kinds of sounds coming from the dark forest that surrounded the *nahs*—frogs, birds, insects. And then toads started to assemble, almost magically, from underneath the platform where Lorenso sat—first one, then more, until there were about two dozen. They were probably attracted to the insects that were drawn to the lights, but it looked as though they'd come to listen to him. This was the theater where stories were told. This was the place where the story lived.

I realized that a translation, no matter how thorough, could not capture it. Still, if I was going to put it down, I felt better that a relative of the teller would be translating for me. As he did, pieces of the story I'd heard previously, tidbits that people had shared, fell into place.

When, days later, we finished the transcription in my room at Yvonne's, Leinson thanked me for coming to Pohnpei to carry out this investigation, for otherwise he would never have heard his uncle's story. I uploaded the digital sound files onto Leinson's computer.

"I will share this with my son," he said, "and he will share it with his children."

As with many indigenous stories, this one is incomplete without a familiarity with the landscape and the creatures that inhabit it. Certain geological features mentioned, such as a large stone ledge at the top of the mountain, can still be seen today, although components of the story are hundreds, if not thousands, of years old. Such stories have practical reasons for existing, acting as maps, with landmarks that punctuate a journey—much like Aboriginal songlines in Australia—or as aids to teach practical knowledge, like the best time of year to fish off a certain reef.

Though the recording of an indigenous story could be considered a failure, I feel that it is valuable to have these tales, if for no other reason than that they are rapidly becoming extinct. With that in mind, here is an approximation, a fragment, a fiction.

Once upon a time on the island of Yap there was a married couple with a daughter, and they had a large *kemisik* [eel] living in the stream near their home. And one day, because the eel was getting big, the couple decided to kill the *kemisik*. But the *kemisik* overheard them, and told the girl that if her parents were to kill him and eat him, she should take his head and put it above the front door of their house, so the head was facing out, and then, after a time, to bury it in the ground.

So when the girl's parents killed the eel, she put

his head over the door, then a few days later she took
it down and buried it, and from the head grew three
things—*mein-iwe*, the breadfruit tree, and two varieties
of banana, *uht mwot* and *uht en yap*.*

On the distant island of Pohnpei, on top of a
mountain, two birds [native Micronesian starlings]
were perched on a ledge overlooking the ocean. The
two starlings were titled Mwahnlaipeip, a male bird,
and Peinlaipeip, a female bird. [The fact that the birds
had titles meant that at one point they were human.]
From their high place in the mountains they could see
something shimmering in the sea. So the male starling,
Mwahnlaipeip, decided to fly out and see what it was.

Mwahnlaipeip flew, flew, flew—but it was too far, he
could not make it, and he returned to Pohnpei. So the
female bird decided to try. She flew out, and flew and
flew and flew, and eventually she saw land. And when
she was close enough, she saw the object that had been
shimmering, the fruit of the banana. [The island she had
flown to was Yap, the place where our story began with
the girl and the eel.]

Peinlaipeip started eating the fruit of the banana,
and while she was eating she swallowed a nut that was

---

* Leinson said that *uht en yap* is a beautiful banana introduced from Yap, with a
red peel and orange flesh. It is often served mashed with coconut milk and is high
in beta-carotene.

inside the banana. [This banana was *uht en yap*, or the
Yap banana, which had grown from the head of the
eel.] After she ate her fill, she flew back to Pohnpei. Just
before she made it to land, she pooped the nut she had
swallowed onto the reef known as Rohn Kitti [part of the
municipality of Kitti].

Meanwhile, two young women from a village called
Kepine were walking down the hillside to take their boat
out fishing on the reef. They stayed out fishing until it
was high tide, then started to come in. On the way back,
they saw waves crashing against something, and when
they went to look, they saw a small stone. One of the
ladies picked it up and was going to return it to the water,
but the other woman asked if she could keep the stone.
[The stone was actually the nut from the banana that the
starling had pooped on the reef.] She put the stone in her
basket and wrapped it in hibiscus leaves. From then on,
this kind of fishing basket was known as *kopwou lasi* [*lasi*
is the old Pohnpeian word for eel].

When the two women had returned to shore, they
began to make their way back up the hill to the village.
Partway up the hill they stopped to rest, and opened up
the basket to look at the stone. When they unwrapped
the hibiscus leaves, they noticed that a crack had
developed in the stone. So they named that place Nan
Ihr, which means "in the crack." They wrapped the
stone up again, put it back in the basket, and continued
walking.

Further up the hill they decided to take another break. At this place, they took the stone out of the basket again and unwrapped it from the hibiscus leaves. To their surprise, the stone had broken apart, and inside they saw what looked like a small worm. They wrapped up the little worm in the hibiscus leaves and continued walking up the steep hill to Kepine.

Eventually the two women reached a river and decided to stop for a drink. While resting by the river they opened up the basket again and unwrapped the hibiscus leaves. Immediately they saw that the little creature had grown, and what they had thought was a worm was actually a baby eel. So they took one of the big clams they had caught on the reef, scooped out the meat, and put some water from the river inside the shell for the eel. If you go to that stream now today, you can see a small pool below the bridge called Douen Lasi, or "place for the eel," referring to the location of this event in the story. Today people call this place Pohn Kati.

From there, the women continued their climb, and finally made it to Kepine. Back home, the woman who had been carrying the eel in her basket took the eel to a small stream beside her and her husband's home and let it go in the water. She took care of the eel and fed it. The eel grew. After many years, the eel had grown very big and they became afraid of it.

One night, lying in bed with her husband, the woman said, "Tomorrow we should kill the eel and cook it."

They planned to start chopping firewood early the next morning to cook the eel.

But the couple didn't know that while they were talking in bed, the *kemisik* was listening to them. By that time, the eel had grown so big it had outgrown the main channel of the stream and made itself a hole under the riverbank beneath their house. There it lived, and from there it could hear everything they were saying, including their plan to kill it. The couple went to sleep.

In Pohnpei, they say, by the third crow of the rooster, the sun has risen, and it is time to get up. But this morning, when the couple woke at the third crow of the rooster, oddly, it was still dark. The husband got out of bed to look outside and see why the rooster was crowing in the dark. But when he tried to go through the doorway, he bumped into something wet and slimy. It was dark because the tail of the giant eel was blocking the door. As the husband looked around, he realized what was happening, and saw that the body of the eel was draped over the rafters above them, and its head was looking down on their bed at his wife.

The husband and wife ran right through the thatched palm walls of their house and up into the forest. They ran and ran and ran up into the mountain without looking back, and hid under a large overhanging rock ledge. But the eel chased them up the mountain, and as the couple hid under the ledge the *kemisik* was on top of the rock, easing her head over the rim. The man and his wife

didn't see the eel and stayed for a while, until it started
to rain. But what they thought was rain was actually
the saliva of the *kemisik* dripping down over the ledge.
They'd backed in as far as they could go and noticed the
rain was slimy. "What kind of rain is this?" they said.
Then they saw the head of the *kemisik*, and when they
tried to run the *kemisik* ate them.

The eel had eaten the woman and her husband, but she
had eaten too much too quickly and couldn't move, so
she lay on top of that rock, not feeling very well. While
the *kemisik* rested on the rock, at that moment Kroumeir,
a spirit who can take the form of a human or an eel,
happened by. Kroumeir was like a god *and* a high chief,
and he was being carried by his people in a cart on their
way to a feast at Nan Madol palace, guests of the ruler
of Pohnpei, the *saudeleur*. In addition to carrying him,
Kroumeir's people were bringing food and gifts for the
ruler.

Kroumeir was fond of the big eel and asked her if
she'd come with them to the feast. She told him to keep
going, that she wasn't feeling well. But Kroumeir didn't
want to leave, so he told his people to go ahead, that he
had another engagement, and to come back for him later.

When the people came back to get Kroumeir after
delivering the gifts to the *saudeleur*, Kroumeir told the
big *kemisik* that he wanted her to come back with him
to his village, and asked her to get in the cart so that she
could be carried there. Kroumeir's people protested.

"Why do we have to carry this ugly, slimy thing, this disgusting animal?" they said angrily. They were afraid of the eel. The word for eel, *kemisik*, comes from the old word for fear, *kamasak*. The people didn't know that Kroumeir himself could become an eel.

Kroumeir placed a *lei,* a ring of flowers, on the eel's head, but her head was flat and he had trouble getting it to stay. Finally he managed to get it to sit on top. That's why the Lasialap people, the Eel Clan, can wear a lei on their forehead and it stays, while most people wear it on the top of the head. The people continued on, carrying the eel (reluctantly) and Kroumeir. Kroumeir was thinking that the eel would be his wife.

When they all returned to the village the people produced a big feast for Kroumeir. During the preparation for the feast and the feast itself, the people complained that Kroumeir had invited this disgusting animal. The eel overheard the people talking about her and told Kroumeir that she felt bad.

"I'm going away for a while," she said, and left the village.

She swam up the Nanpil River, over the mountain, and down the Rohn Kitti River to the other side of the island. The headwaters of these two rivers of Pohnpei are very close, and it is possible even today that an eel could swim up and over the mountain from one side of the island to another. The eel continued out the Rohn Kitti to the ocean, where she saw some people fishing on the reef.

It was in March, during the season when many fishes were congregating off the reefs to spawn. Two boys were fishing at the edge of a big school of snappers. The eel asked the boys if she could fish with them. So they let her fish.

The big eel opened up her mouth, sucking in one school of fish, swallowing it whole. Then she swallowed a big wave that was crashing on the reef to wash the fish down. The eel then opened her mouth again and swallowed a second school of fish, and then another wave. Then she swallowed a third school of fish, and another wave—then a fourth school of fish, but not a fourth wave. She swallowed four schools of fish in all, and three waves, and thanked the two boys and left the reef.

She swam back to land and up the Rohn Kitti River. On her way upstream, she ran into an eel god named Kroumand, who was hunting for birds in the jungle in the form of a man. Kroumand had killed so many birds that he had them hanging from his ears and a belt of birds around his waist. The eel got scared of Kroumand and hid from him. Not knowing she was there, Kroumand stepped on her. She was pregnant at that time with Kroumeir's child, and when Kroumand stepped on her she gave birth to the first Lasialap. The place where she gave birth, they call that place Lipwentiak—*lipwen* means "print," and *tiak* "foot," so Lipwentiak is the place of the footprint. Further upstream the eel gave birth again, but that baby did not make it. The unborn baby became a tree—*lasi o dong*. *Lasi o dong* is a large endemic tree that grows in the mountains and has a fruit called *lasi kotopw*. [Mauricio, the island historian, had said that Lasi O Dong and Lasi Kotopw were subclans of the Eel Clan.]

The eel continued up over the mountain and down the Nanpil River to the village again. Kroumeir could sense that she was coming and said to his people, "Make a feast! My wife is returning from fishing." The villagers complained. They were especially annoyed when the eel arrived, because Kroumeir said she had just come from fishing, yet she did not carry anything to contribute to the feast.

The eel asked Kroumeir to tell his people to bring banana leaves and spread them out around the *nahs*, the traditional meeting place. The people followed Kroumeir's command and brought bundles of broad and shiny banana leaves. The eel said to bring more. The villagers brought more banana leaves. The eel said to bring more still, and then to bring more again. When they had laid out the fourth area of banana leaves she said, "That's enough." That's why, today, when there is a party at a *nahs*, before you spread out a feast you start with four banana or palm leaves.

The big eel told everyone to stand back. She started to gag, like she was coughing, and threw up a school of fish all over the banana leaves laid out in the *nahs*, and then she threw up a wave of water to clean it. Then she threw up a second school of fish and another wave to rinse those fish—then a third school of fish and a third wave. Then the fourth school of fish came but no wave of water followed. That's because, if you remember, she only swallowed three waves of water. And that is why, to this day, parrotfish (the fourth type of fish) are so slimy behind the gills, because they came from the slimy eel and were not washed off with a wave of water. This is the species of parrotfish that the locals call *mahu*, which is especially slimy, the very blue one.

Once the eel had delivered her gift, she left the party and swam up the Nanpil River and over the mountain to Lehdau, in Madolenihmw. She stayed in the village of Sapalap for quite a while, living in a tidal channel called Dau Sokele. Every time a canoe went out across the channel, the giant eel asked for one fisherman to be dropped off the boat. And so she ate the fishermen in Sapalap one at a time as they went out of the channel to go fishing on the reef. And very soon those people realized that if this continued, none of them would be left. So they came up with an idea to trick the eel.

They filled a canoe with coconuts, palm fronds, and even big shells that would make noise as the wind blew through them. It was like a party on the boat. And they put a sail on the canoe and pushed it out of Dau Sokele channel toward the island of Kosrae. The canoe sailed outside of Tomwhak pass and beyond the reef to the sea.

When the eel was very close to Pingelap, she realized something was wrong. "Who are these disrespectful people, not dropping off a fisherman for me?" she asked. She continued following the canoe all the way to Kosrae, until finally she overturned the canoe and discovered the trick. "There's no people in this canoe. It's full of coconuts!" So she ate all the coconuts and the canoe and turned back to Pohnpei.

Unfortunately for the eel, as she was returning she was attacked by a shark. So when she finally reached Pohnpei and entered the Nett pass, there was only half of her left. She died on a piece of the shoreline. And that whole piece of land that is left over in Nett is her body.

And that is the end of the story. *Ahi soai pwoat torohr wei likin imwen.* Pass it on—from this house to people outside of your house.

The day before I left Pohnpei, I met with others who told pieces of the story that fit into the one related to me by Lorenso Gilmete. It seemed that unless I was the object of an island-wide ruse, the basic skeleton of the story was consistent.

I met with Paul Gallen, a high-titled Protestant minister and the uncle of the night watchman at Yvonne's. He had the most detailed version of the part of the story where the villagers trick the eel with a boat full of coconuts.

And then I met with another man near Awak, a high-titled Lasialap named Mikel Marquez. This opportunity came about because his son, Roseo, worked for CSP. In order to sit with

Mikel Marquez we had to have a *sakau* ceremony. Mikel had been in poor health and suffered a series of minor strokes; he said politely that he had once known the story but could not remember it now.

"You should have come two or three years ago," Roseo said to me openly, in front of his father. Roseo went off to clean the *sakau* root, and I was left alone with his father.

We were seated in the *nahs* surrounded by a lush and verdant garden, and Mikel said, "There is an eel in the stream. Do you want to see him?" Mikel got up and walked slowly to a tiny creek that trickled through his garden. Mostly concealed by a massive stone slab was a large eel, but the water was barely deep enough to cover the eel's body. A boy came down to the creek, seemingly from nowhere. Mikel had a word with him in Pohnpeian, and the boy touched the eel's head. All of a sudden the eel came out of its hole under the rock slab, completely exposing its body as if it wished to be petted.

The boy slipped his hands under the eel and lifted it up out of the water, with no protest from the fish—it actually appeared the eel liked being held by the boy. Mikel told me that this was the fifth eel that had lived here in his lifetime. "When one gets old it moves away, and a smaller one comes in to take its place," he said. "One day my grandson was poking the eel with a stick, and the eel chased him all the way back to the *nahs*, nearly fifty feet. Luckily I was there to see it. I scolded the eel, and it went back to its place."

When Roseo and his friend Kesdy started pounding the *sakau*, they invited me to participate. I followed suit and took off my shirt, selecting a rounded stone from a box full of pounding stones near one of the posts of the *nahs*. It was satisfying to smash the *sakau* with the stone and create the hollow metallic sound that I'd first heard in the jungle. When we had sufficiently mashed the *sakau*, two younger men came by to squeeze it and present the drink. They were official *sakau* servers, designated to prepare the drink for high chiefs.

The first cup was handed to Mikel. The second squeeze went to the person of second-highest rank, his son, Roseo. The third cup was offered to Mikel's wife, and the fourth cup was handed back to the chief, who passed it to me, which, Roseo explained, "means that he feels you are a person of importance." Humbled, I drank the slimy liquid.

The sky grew darker and the sakau began to take effect. We stopped speaking, and the big eel in the stream started to splash loudly. No one acknowledged verbally that the eel was splashing in concert with our silence, but everyone seemed to be focusing on the sound of the splashing. Roseo was the first to break the silence. "Up to this day," he said, "for the Lasialap, the eel is their totem. They protect and worship the eel, they look after the eel."

We sat silently for some time again, and the eel resumed its splashing in the creek.

"I used to believe I was born from eels," Mikel said, "but now not as much."

I asked why.

"I think it's the religion," he said, meaning the introduction of Christianity. "Have you ever heard the eels?" I shook my head. "They make a sneeze, or a whistle—a melody just like a bird. I've seen, with my own eyes, eels walking on their tails on land. Not too fast, not too slow, just standing on their tails. They climb trees to get birds." He lifted his hand to demonstrate. "I saw an eel going after a fairy tern in a tree ten feet off the ground. That used to happen, but no more. That eel is no more."

As the night wore on and we drank more *sakau*, Mikel asked if maybe I could tell the story of the *kemisik* as I'd heard it so far on my trip. So I did, with my own little flourishes, as best I could remember.

"On the island of Yap," I began, "there lived a couple who had a daughter, and in the stream near their home lived an eel. The couple fed the eel and one day it got too big and they decided to kill the eel. The eel overheard the couple and told the girl, 'When your parents cut off my head, put it over the door to your home facing out, and then bury it in the ground.' She was fond of the eel, and did so, and from the head of the eel grew breadfruit and two kinds of banana."

I continued, telling how the starling saw the glimmering light in the sea and flew for it and ate the seed of the banana, and brought the seed back to Pohnpei, which became the eel, and how that eel ate the couple that had raised it, and then met the god Kroumeir and had various children that became the first Lasialap people. And the whole time I was speaking, I saw that Mikel was nodding and smiling. I was surprised how much I remembered.

I paused in the middle and looked at Roseo. "My father is

nodding," Roseo said, "because you are right. You are helping him remember the story."

Everyone was quiet now from the *sakau*.

As I went on, Mikel continued his slow nodding. A few times he even stopped to correct me or to help me to pronounce a name. He seemed now to remember quite a bit, as if the story had been blocked and was released by a little priming.

Mikel added one final note. "When the eel was bitten by the shark," he said, "she was near Pingelap, and the eel took sand from Pingelap to cover her wound. And when the eel, tired and nearly dead, arrived back in Pohnpei, she put the sand from her wound on the ground beneath her body. That's why the shore there is called Pingelap. The shark's name was Nahn Sou Set."

I did not know if it had been intentional, but when I finished the story, I realized that the roles had been reversed—that Mikel had turned me into the storyteller when it was I who had been seeking the stories in the first place.

chapter ten

Obstacles
IN THEIR PATH

Hydropower turbine

In the late 1990s, John Casselman, a biologist at Queen's University in Ontario, began to expose an alarming decline in the number of juvenile eels returning to the St. Lawrence River in Canada, at the northern edge of the range of the American eel. The population of young eels coming up the fish ladder at the Moses-Saunders hydroelectric dam had dropped from nearly a million in the 1980s to a hundred thousand in the 1990s, to less than ten thousand in the late 1990s and virtually to zero in 2000. Casselman stated that, historically, female eels once had made up 50 percent of the inshore fish biomass of Lake Ontario at the head of the St. Lawrence; now, almost no eels were returning there at all.

"It is truly a crisis," Casselman told me.

Even more alarming is that what Casselman was observing was not unique to the rivers of eastern North America. Populations of freshwater eels the world over were in serious decline. The *American Fisheries Society Journal* stated in 2003: "In recent decades, juvenile abundance has declined dramatically: by 99% for the European eel, *Anguilla anguilla*, and by 80% for the Japanese eel, *Anguilla japonica*. Recruitment of the American eel, *Anguilla rostrata*, to Lake Ontario, near the species' northern limit, has virtually ceased."

There was no way around the fact that for a migratory fish such as the eel, the existence of hydropower dams was a major issue, perhaps *the* major issue, contributing to the species' decline. Construction of the Beauharnois and Moses-Saunders hydroelectric dams on the St. Lawrence in 1932 and 1958, respectively, had impeded the migrations of eels to and from what once comprised the single largest nursery in North America—the upper St. Lawrence River, Lake Ontario, and their feeders.* Even if a juvenile eel was able to make it upstream of the dams via fish ladders, the downstream gauntlet was nearly insurmountable.

During the fall migration the accumulated mortality from the turbines of both dams was about 40 percent, Casselman told me, but that didn't account for the fish that were wounded and weren't in good enough physical condition to make the long journey to the Sargasso. "Unfortunately," Casselman added, "since the dams on the St. Lawrence are run-of-the-river and use all the flow, there is little or no likelihood that the eels can get by any other way."

In April 2000, the Atlantic States Marine Fisheries Commission (ASMFC) issued an extensive Interstate Fishery Management Plan for the American eel, recommending necessary steps to save the species. Despite further evidence in the report of a "very serious" decline along the eastern seaboard, the proposed plan was never put into effect. In March 2004, the ASMFC Eel Management Board issued a press release, recommending that the United States Fish and Wildlife Service (USFWS) consider protection of the American eel under the Endangered Species Act (ESA). The USFWS failed to respond.

The evidence of a drastic decline was accumulating fast and becoming irrefutable. There were no longer enough eels being born in the Sargasso Sea to spread to the extremes of the fish's

---

* The St. Lawrence River, which drains the Great Lakes, contains about 17 percent of the freshwater habitat within the range of the North American eel.

historical range. There were still plenty of eels in South Carolina, Virginia, and Maryland, closer to the spawning grounds, but in places such as the St. Lawrence in Canada or the Mississippi, where eels were once abundant, they were hardly showing up at all anymore. Likewise, in Europe, populations were still somewhat healthy in rivers of France and England closer to the eels' birthplace in the Sargasso, but beyond the strait of Denmark in the Baltic Sea or at the eastern end of the Mediterranean Sea (and up the Nile) there were far fewer. The European Union threatened to shut down the commercial fishery for eels in Europe, period. Pressure was mounting in the United States to do something as well.

Because of their sliminess and association with the snake and the phallus, the eel has a tendency to stir human uneasiness, even disgust. It is not, like the salmon, swordfish, or giant tuna, an easily embraced, bold, and magnificent emblem of the rivers and seas, attracting funds to the coffers of foundations. Who would step forward, then, on behalf of the eel? Maybe it would take someone a little different and tenacious, someone who moved in the opposite direction of everyone else, to become the spokesperson for the eel.

And so it was that Tim Watts, a janitor at Stonehill College in Easton, Massachusetts, was combing the Atlantic States Marine Fisheries Commission website one day when, out of personal curiosity, he began reading the minutes from meetings of the ASMFC's Eel Management Board.* He read papers delivered by John Casselman and other scientists and technicians, as well as testimony from amateur naturalists and professional fishermen. All the data and personal commentaries pointed to the fact that the eel population was nose-diving up and down the East Coast.

About the same time, in the autumn of 2004, Tim's brother, Doug, a journalist living in Augusta, Maine, started noticing hundreds of dead eels showing up below the Benton Falls dam on the Sebasticook River, a major tributary of the Kennebec. On the phone one night, the brothers discovered that their observations on both the national and local levels converged. They felt they had to do something.

Unable to contain their anger at the U.S. government's lack of action regarding the eels' decline, in December 2004 the two brothers submitted a citizens' petition to list the American eel as an endangered species under the Endangered Species Act. Because of the thoroughness of their research and the irrefutable evidence in their petition, the U.S. government was forced to take their plea seriously, promising to carry out a full-blown

---

* At the March 29, 2004, meeting of the American Eel Management Board of the Atlantic States Marine Fisheries Commission, Eric Smith, the commission's Connecticut representative, stated: "I just want to see that the issue doesn't languish until such time as five years from now, we say, okay, now we've got some landings [data] and can deal with it. This thing has been troubling for some time now that you look at the slides."

status review of the species. It was a triumph for the brothers, for blue-collar America, and for the eel.

The brothers' passion for the eel made for a compelling story and garnered national media attention. The *New York Times* and *USA Today* as well as several National Public Radio programs ran pieces about the brothers and the eels' plight. Doug and Tim Watts's action demonstrated that individuals had the ability to make a difference. With pen and paper, two ordinary citizens had forced the United States government, at the very least, to listen.

Soon after reading in the *New York Times* of the proposed listing of the eel under the ESA, I tried contacting Doug Watts. I was given his home number in Augusta by a friend who'd worked with him on the ESA listing of Atlantic salmon in Maine. He warned that Doug could be somewhat unpredictable and was notorious locally as a fist-pounding activist who sometimes did outrageous things to make his point.

"Is this Doug Watts?" I asked when I called.

"Yes," he answered in a low, gravelly voice.

"I'm a writer working on—"

Click.

I called back, thinking we'd had a bad connection, but no one answered. I called about a dozen more times. Nothing. I left him alone. Maybe he was getting threats from eel fishermen.

Unable to get to Doug in person, I began to follow the work of the Watts brothers through their website, Glooskapandthe Frog.org (Glooskap is a hero in Penobscot Indian creation myth). One post showed a provocative and, frankly, odd picture of curly-haired Doug Watts with no shirt on, holding in front of his face an eel that had been killed in a hydropower turbine in one of Maine's dams. Behind him was a picture of the folk music legend Blind Lemon Jefferson. The post was titled, "What Is Right and Wrong?"

It's frustrating, if this slaughter involved monarch butterflies, roseate terns or the sea turtles . . . the public outcry would bring turbine blades to halt and rattle to rubble a thousand dams that haven't powered a mill for half a century.

Because fish are being slaughtered, slimy eels, we minimize and justify the slaughter as the price to pay for Green Power.

Dams on our rivers, especially our co[a]stal ones are to our rivers and their aquatic ecosystems are what pesticides once were to our ecosystems on land. The difference is awareness, as of yet there has been no Rachel Carson to write the story and challenge the status quo.

In addition to shocking pictures of dead eels, quotes by Henry David Thoreau, and the arresting phrase "Complacency Kills!" in bold type on the home page, the website had beautiful photographs taken by Tim Watts of glass eels and elvers successfully scaling the wet wall of a bridge abutment to get above a defunct dam. The power of the eels' resilience to overcome the obstruction was inspirational, as were the love and patience Tim must have had to document this segment of their long journey.

If you look at a map of eastern North America and imagine that the rivers coursing through the land are the blood vessels of a human body, that body most certainly would be experiencing cardiac arrest. Few rivers flowing east from the Continental Divide to the Atlantic Ocean are now free of dams. The health of the earth's circulatory system, a network of interconnected ecosystems—from springs and swamps to rivers and ocean currents—depends significantly on a free exchange of organic and inorganic matter.

The effects of dams on the ecosystem are not always easily seen. For example, the entire length of the Susquehanna River used to be thick with freshwater mussels, which naturally filtered the water and provided food for numerous creatures. Biologists at the U.S. Geological Survey, after completing a survey of freshwater mussels on the Delaware River, which essentially has no dams and supports about two million mussels per mile, wondered why freshwater mussels were now virtually nonexistent in the nearby Susquehanna basin. The researchers found that the larvae of freshwater mussels attach themselves to hosts that help distribute them up and down the river before the larvae drop off and become small mussels. It happens that the most common species of mussel, *Elliptio complanata*, prefers to attach itself to a particular species of fish known for migrating nearly the entire length of the river system—the freshwater eel. In the early 1900s, a series of large hydroelectric dams was built in the lower reaches of the Susquehanna, preventing eels from populating the upper reaches of the river.* Without the eel, the freshwater mussel lost its host distribution system and disappeared.

One might ask, what were the consequences for the Susquehanna ecosystem of losing the eels and freshwater mussels? It is thought that eels once accounted for 25 percent of the total fish biomass of East Coast rivers and streams. Certainly that biomass was now no longer available to feed other creatures, from ospreys to raccoons, herons, and striped bass. But there were other consequences. In a healthy system such as the Delaware, the mussels were filtering more than two billion gallons of water per mile each day. Without the mussels filtering the Susquehanna, how was the river's general health being affected?

Throughout much of their native range Pacific salmon can no longer reach the headwaters of the great rivers to spawn; neither can shad, alewives, or other migratory fishes. It is not only the salmon's existence that is at stake, but the existence of every

---

* The Maryland Department of Natural Resources *MBSS Newsletter*, March 1999, states: "The most dramatic example of the decline of American eel abundance is dam construction on the Susquehanna River. Prior to the completion of Conowingo and three other mainstem dams in the 1920s, eels were common throughout the Susquehanna basin and were popular with anglers. To estimate the number of eels lost as a result of construction of Conowingo Dam, we used MBSS data on American eels from the Lower Susquehanna basin and extrapolated it to the rest of the basin above the dam. Our best conservative guess is that there are on the order of 11 million fewer eels in the Susquehanna basin today than in the 1920s.

"Because adult eels migrate to the Sargasso Sea to spawn and die—transporting their accumulated biomass and nutrient load out of Chesapeake Bay—the loss of eels has increased nutrient loads in the basin and reduced them in the open ocean where they are more appreciated."

creature that their stray eggs and post-spawn carcasses once supported—the caddis flies, the grayling, the rainbow trout, the eagles, the bears. What is at stake if we lose the eel?

"The eels may be sending us an important message," John Casselman told me. "The question is—are we reading it?"

⁓

A few years after I first called Doug Watts, I tried him again, this time by e-mail. By chance, he had seen an article I'd written for a fly-fishing magazine about the conservation of native trout.

"I saw your piece," Doug responded. "I'm kind of disenchanted with fly fishermen these days. I think for lots of them trout are golf balls that swim. Your comment to that effect in your piece was spot-on and much appreciated."

He said I was welcome to come up and visit him in Augusta, Maine. We set a date, and I drove up.

Doug lived with his wife in a modest home in the industrial outskirts of town. She was a potter, and Doug had his hand in other arts and music besides his sporadic freelance writing gigs and activism.

We sat down across from each other at a table in the kitchen, beside a few potted spider plants and African violets. Doug had wild brown eyes, a lean body, and curly brown hair. He gave the general impression of a fugitive, pulsing with nervous energy. One of the many cats milling around jumped on his lap. With unwavering intensity, Doug told me what had pushed him and his brother over the edge.

"The Sebasticook comes into the Kennebec about twenty

miles above Augusta, and Benton Falls is the second dam on that river," he began. Benton Falls is a hydrodam. It was constructed in 1987, the last new hydrodam ever built in Maine. Doug had already known from reading Maine Department of Natural Resources reports that this particular dam was slaughtering big female eels. These were eels that had gone up the Sebasticook to the headwater lakes as juveniles, before the dam had been built, and lived there until their hormones clicked and they were ready to return to the sea to spawn. "Only now," Doug said, "they couldn't. You're talking about thirty-, forty-year-old animals. I've handled hundreds of eels maimed by hydrodams, whose skin has been pulled off like a sock off your foot. I tell people it's like sticking your hand in a window fan, except the fan's much bigger and the metal's much stronger."

Doug is not gentle with his metaphors or visuals. His website shows other pictures of mutilated eels, one with the following caption: "Those are the eel's gills hanging out of her mouth. The force of the turbine blade strike blew them out of her throat. Picture yourself getting hit in the abdomen with enough force to spit your lungs up."

Doug went down to the dam and hunted around in the water. The river bottom at the base of the dam was covered with eel carcasses. Some of them were dead, in pieces; some weren't dead yet, just maimed and swimming in circles. Doug called the state and told them what he'd seen. The state said the power company that operated the dam was not doing anything illegal—the dam wasn't breaking any laws. But they asked the power company to voluntarily shut off the turbines. The dam owner, down in Atlanta, Georgia, said, in essence, "Screw it—we don't care."

"There were hundreds and hundreds and hundreds of eels all over the bottom of the stream," Doug said. "The state wouldn't do anything. The dam owner wouldn't do anything. And so Timmy and I talked on the phone. And, you know, we were *livid*. We were just livid to the point of wanting to go do violence to things. Mostly because the state of Maine just threw up their hands and said, 'We don't care.' "

What happened next was the kind of activism that had made Doug Watts notorious in the state of Maine. He walked right off the street into the Maine Department of Environmental Protection offices in Augusta with a plastic grocery bag containing five huge dead eels. "You don't believe me? Here you go," he said to the officials, dumping the carcasses on their carpet. For Doug it was all-out war.

After repeated efforts, Doug was getting nowhere with the state. Tim pointed him to the ASMFC website and John Casselman's research on the decline of eels in the St. Lawrence River. He said to Doug, "If this guy Casselman is right, it looks like the whole population is starting to go down." Telling the story to me, Doug raised his voice angrily. "You can only beat an animal so hard before it finally just gives up. You can only subject them to so much cumulative abuse before literally their back breaks and the population just crashes."

Doug offered the passenger pigeon as a very useful analogy. In the nineteenth century, John James Audubon wrote of clouds of migrating birds that numbered in the billions. The population of passenger pigeons was estimated to be over six billion, representing 25 percent or more of the total bird biomass of North America. Even though millions of passenger pigeons were being killed every year, it seemed the resource could never be exhausted.*

"And that's what they also said about the Newfoundland cod," Doug continued. " 'There's no way you can—' " Doug cleared his throat and raised his voice even louder: *" 'There's no way you can fish out cod!' "* He repeated himself more slowly, and even louder: " 'There's no way you can fish out the Grand Banks cod population off Newfoundland! That's insane! That's like saying you can catch all the plankton! Like you can catch all the salt in the water.' And guess what? There are no passenger pigeons anymore—we have confirmation of that. And there's no Newfoundland cod fishery—it's gone, it's done."

Similarly, it was once said that the superabundant eel population of the St. Lawrence could never be diminished. In an article in the *New York Times* on October 18, 1880, the author wrote, "The eel fisheries of the St. Lawrence are probably the most

---

* Edward Howe Forebush wrote in the early 1900s in *A History of the Game Birds, Wild-Fowl and Shore Birds of Massachusetts and Adjacent States*, "No adequate attempt to protect them [the passenger pigeon] was made until they practically had disappeared. Whenever a law looking toward the conservation of these birds was proposed in any State, its opponents argued before the legislative committees that the Pigeons 'needed no protection;' that their numbers were so vast, and that they ranged over such a great extent of country, that they were simply able to take care of themselves. This argument defeated all measures that might have given adequate protection to this species. That is why extinction finally came quickly. We did our best to exterminate both old and young, and we succeeded."

productive in the world, and the quality of its eels is considered to be unsurpassed. This mysterious fish seems to hold its own against the demands of civilization better than most others. It seems likely . . . that this valuable supply of food will remain abundant here for an indefinite period." But in the early years of the twenty-first century, warning shots were being fired; the eels were disappearing.

"And so," Doug said, "that's why we filed the petition. Because the ASMFC voted to send a letter to the U.S. Fish and Wildlife Service, asking them to do a status review under the Endangered Species Act. We're like, 'Hey, if the ASMFC is actually seriously talking about eels getting listed, it's like, holy crap, this is bad!'

"But we knew nothing would happen," Doug added. "I said to Timmy, 'The ASMFC isn't going to submit a formal petition to list, and without a formal petition it isn't going to trigger the process. But hey, *we* can.'"

So they did. They filed their petition in November 2004. And shortly thereafter, a journalist from the Associated Press wrote a story on the brothers that ended up running in major newspapers across the United States.

Being a reporter himself, Doug admitted he could see what was happening. "The reason this is getting coverage," he said, "is because the lead is so good: 'A *Massachusetts janitor* and his brother, an *unemployed writer*, have petitioned the federal government to list the American eel, a *slimy, icky thing that no one cares about*, as an endangered species.' It's such a ridiculous story, they put it on the wire. 'The two brothers want to protect probably the most *unlikable* animal in the world, next to, I don't know, hair lice.'" He cleared his throat. "'A *Massachusetts janitor* is concerned about the welfare of the *tick*!'"

Doug got up to stretch his legs and make some coffee, shaking his head and laughing. I asked him how one went about the process of writing a "petition to list." He answered me with a smirk in his thick eastern Massachusetts accent.

"I went on the Internet," he said, putting water in the coffeepot, "and typed in a search for 'petition to list Endangered Species Act.' I looked at a bunch of other petitions and saw the legal format that they used. So all I did was, every time I saw 'Alaskan sea otter' I crossed it out and put 'American eel.' Total cut-and-paste." When Doug saw I was falling for his joke, he broke into uproarious laughter.

Actually, Doug was quite learned about the process, having been involved in listing the Atlantic salmon as an endangered species in Maine. When a group sued to force the listing of salmon, he was a plaintiff in that lawsuit. He wrote the eel petition in a week, Tim having done most of the research. It is thirty-five pages of very readable and seemingly endless historical and contemporary evidence, all pointing to the same sad fact.

"Once the formal petition is submitted to the government," Doug explained, "it starts a statutory clock. The ESA has ninety days to decide if the petition has enough scientific backing to warrant a full-blown status review of the species." Doug said it took the government six months to complete the ninety-day review. But in July 2005, the USFWS published the ninety-day notice, determining that the Watts brothers indeed had presented enough scientific information to warrant a full status review.

"It was very empowering," Doug said. "It was like, 'Oh my God, I can literally sit at my computer and write up this document that can force the government to do something. All this for a forty-cent stamp!' "

Eels are native to the entire Atlantic seaboard. They also historically went all the way up the Mississippi and its tributaries—the Ohio, the Tennessee. At one point eels were essentially native to about two-thirds of the continental United States. Doug realized that because the eels' range was so huge, a listing of eels under the ESA would affect every single facility that puts water into or takes water out of every river or stream east of the Rockies. Doug said to his brother, "You know something, Timmy? This is the H-bomb of listings."

Doug got up and refilled our coffee mugs, adding a few heaping spoons of sugar to his. "Timmy and I, we knew the fix was in, because if the service ever listed eels, even as a threatened species, which is a lower protective threshold, it would affect so many places, so many people, so many businesses, so many industries. We were like, 'Yeah, right. This ain't goin' nowhere.' "

Once the petition to list is submitted and the government determines that a status review is warranted, it has one year to make a decision about the listing. In the case of the Wattses' eel petition, eighteen months went by and there was still no word from the government. Tim kept in touch with Heather Bell, the biologist at the USFWS assigned to assemble and write the status review, asking her when they might have a decision. She said she didn't know. A few more months went by. They had just about given up on getting a response from the government at all. Then, out of the blue, a law firm in Washington, D.C., that specialized in ESA litigation contacted the Watts brothers to take on their eel case, pro bono.

Meyer and Glitzenstein sued the USFWS for its tardiness. That lawsuit dragged out for six months. In January 2007, the USFWS finally promised it would come out with a decision by February.* "If we had not sued them in federal court," Doug said, "they never would have come out with a decision. There was no need to; why bother?"

On February 2, 2007, the USFWS announced in a thirty-page report, written by Heather Bell, that listing American eels under the Endangered Species Act was "not warranted."

"While the eel population has declined in some areas," Bell wrote, "the eel population as a whole shows significant resiliency. If we look at eels over time, we see fluctuations in the population numbers, so a decreasing number of eels right now does not necessarily forecast an irreversible trend. The species'

---

* The law firm's pro bono work for the Watts brothers was eventually paid for by the U.S. government when it was determined that USFWS was in default of its legal responsibility.

overall population is not in danger of extinction or likely to become so in the foreseeable future."

"The conclusion itself is insane!" Doug said. "I've read tons of this stuff. This one, I could tell they cooked the books."

In standard status reviews, they identify all the different things that are negatively affecting the species under investigation, from birth until death. And then, at the end of the report, they consider how all these effects together are contributing to the potential extinction of the species. This is called a cumulative effects analysis. In the USFWS final report, Doug said, they looked at all the factors affecting the eel separately, in isolation, but never considered them together.

"In the real world," Doug said, "eels are being affected by all this stuff *at once*—toxics, dams, fishing, habitat degradation. Perhaps climate change, disease—there are even some nasty nematode parasites spreading through the population. What they never did was ask, 'Might all these factors together be a threat to the continued existence of the freshwater eel?' You know, I read over the thing like ten times, and I said, 'There's something missing here.'"

A cumulative effects analysis tries to predict what all those problems can do together—as Doug calls it, "death by a thousand paper cuts." This is routinely addressed in standard reviews, such as those for the Atlantic salmon or the Atlantic sturgeon, and is one of the most basic components of an ecological report. It was also clear to Doug, from the length of the report and the breadth of the research cited, that the government hadn't done its homework, or didn't want to.

"The status review for Atlantic salmon was three hundred pages," Doug said. "The one for eels was thirty."

Doug called Bell, the author of the eel report, and held her accountable.

"I said to her, 'This is Ecology 101. If you submitted this as a master's thesis, you'd fail!' I said, 'Heather, do you *honestly* think you could publish this in the *American Fisheries Society Journal*? You *couldn't*, they'd reject it. You're putting out an official USFWS legal finding that you know would be rejected in a peer-reviewed scientific journal. Would you honestly submit this?' No, she wouldn't, because an attorney wrote it. It wasn't a scientific document; it was a *legal* document. I said to Heather, 'I know this isn't you; you're smart. This is you under duress.'"

Besides not having a cumulative effects analysis, the report made an audacious assumption about the life cycle of the freshwater eel that absolved the dams of their effect on the species. The findings basically said that eels don't need freshwater habitat to survive. Doug threw up his hands in exasperation. "That's like saying bald eagles don't need trees to nest in! They can use telephone poles."

The status review uses a paper written by Katsumi Tsukamoto to support this argument. Katsumi's paper basically says that yes, eels *can* complete their full life cycle without entering freshwater, that eels are not a true catadromous fish but have a choice of whether or not to enter freshwater. This behavior is called facultative catadromy.

"The eel is a very plastic species," Katsumi said. "You can tell by strontium and calcium deposits in the fish whether or not they've been in freshwater. And we found that some never go up the rivers." This phenomenon of eels living in estuarine waters of various salinities was later documented by scientists for all the northern temperate species of eels, including the American eel and also the New Zealand eel.

Time will tell whether eels will be able to adapt to these new conditions. They are certainly resilient and have, after all, survived ice ages, when major parts of their northern freshwater habitats were covered with ice for millennia. There is no arguing, though, that the larger the amount of habitat and rearing area eliminated from the eels' historic range, the fewer eels there will be.

In fairness to Heather Bell at the USFWS, she did her job thoroughly.

Mike Miller at the Ocean Research Institute in Tokyo argued, "The government under the ESA is not evaluating whether the eel is in severe decline, or in trouble, or getting killed by the millions, they are trying to determine if it is threatened with extinction. Now, that may be more a flaw in the system—that there is no organization taking preemptive measures to protect species in steep decline."

The USFWS held at least four major workshops that assembled scientific experts from many fields of eel biology and ecology to determine and discuss the present state of knowledge about the American eel. They concluded that there is no evidence of possible extinction. Population size reduction is irrefutable, however. Dams are likely a big factor, but Miller, who was flown to the States by the government to participate in the workshops, told me, "More evidence is building that the problem may be in the ocean. The timing of the decline does not coincide well with the building of most of the dams in North America, but they do coincide better with changes in the ocean atmosphere system, it seems."

John Casselman's primary concern was not that eels would become extinct, but something more abstract.

"What I fear," Casselman said to me, "is not that we'll lose the eel completely, but that we will lose our association with them. The eel fishery was highly productive for our First Nation people like the Iroquois, and enabled them to survive when all else failed."

chapter eleven

STILL IN THE

Hunt

The eel

Curiosity will never be content. Even today, when we know so much, curiosity has not unravelled the riddle of the birth and sex life of the eel. Perhaps these are things, like *many others, destined never to be learnt before the world comes to an end. Or perhaps—but here I speculate, here my own curiosity leads me by the nose—the world is so arranged that when all things are learnt, when curiosity is exhausted (so, long live curiosity), that is when the world shall have come to its end. But even if we learn how, and what and where and when, will we ever know why? Why, why?*

—GRAHAM SWIFT, *WATERLAND*

In the fall of 2008, tying up the loose ends of this book, I e-mailed Mike Miller at the Ocean Research Institute in Tokyo to ask a few last questions. Mike said he had some big news but could not share it until the official press release had gone out to the Japanese media. I had a feeling what his news might be, and though I didn't know the details, there was really nothing else it could be: they had found them.

Some weeks later Mike told me that a vessel operated by the Japanese Fisheries Research Agency had, for the first time, caught adult specimens of a catadromous eel on the spawning grounds. The vessel had netted single adult males of both the Japanese eel and the giant mottled eel, *Anguilla marmorata*, and two female specimens of *Anguilla japonica* in professional fisheries' trawling nets not far from the area where he and Katsumi Tsukamoto had captured the smallest newborn eel larvae to the west of Guam. "The discovery," Mike wrote me in an e-mail, "was not that glamorous," in part because the eels caught had completed their spawning, and were flaccid and nearly dead, except for one male *Anguilla marmorata* that was still alive enough to swim in a tank on board for a while.

"But adults have finally been caught in the ocean, at least," he added with little fanfare.

I called Mike early one morning in Tokyo to get a read on what the summer discovery might actually mean to him and his colleagues. I told him that I had feared this moment. Mike downplayed the discovery.

"Finding the adults doesn't really tell us anything we didn't already know," he said. "The mystery is still there. We still don't know how they do it. We don't know how many there are, whether they arrive at once or spawn in stages. We don't know exactly where they do it." As he explained, the eels caught had spawned days before and possibly drifted a good distance from the spawning area. "And we haven't seen them doing it."

But with each year, researchers peer more into the hidden lives of eels. The Europeans are accelerating their eel tagging projects, just as they are facing an alarming crash in the return of juvenile eels to their rivers. In the autumn of 2006 and again in 2008, scientists released adult eels from the west coast of Ireland outfitted with pop-up tags to try to track them out to the Sargasso Sea.

"Do you think they'll find them?" I asked Mike.

"Well," he said matter-of-factly, "there's nothing really to find. We know where they spawn. That's not a big question."*

---

* In November 2006, researchers released twenty-two adult eels outfitted with tags from the west coast of Ireland. The tags were timed to pop off the eels in April 2007, float to the surface, and transmit data on the eels' travels via satellite

I asked why the Fisheries Research Agency of Japan had spent so much money to find the Japanese eels. What value was it to them?

"The agency wanted to know about the spawning ecology, where the eels spawn exactly, and how many are spawning," Mike said. "They sold it from a fisheries management perspective. But basically they were just cowboys wanting to get out there and see if they could catch them." Finding them did not add much to our scientific knowledge of this fish, Mike suggested, nor did it take anything away from the magnificence of their life history.

"Such a long migration as that made by the European eel and other catadromous eels is rare among fishes," Mike mused. "It's *incredible*. I mean, giant tuna move around the globe pretty far, too, and come back to spawning areas, but they're not coming from freshwater and doing it. And there's very few species that migrate that far just once in their lives and then die."

Mike thinks the position of the spawning area is imprinted on the eel larvae when they're born, by some geomagnetic map sense.

"It has been proven that eels have a magnetic sense," Mike said. "And I hypothesize that the reason they have that magnetic

---

to a computer. But in the five months the researchers predicted it would take the eels to get to the Sargasso Sea, they had covered only 800 miles of the 3,000-mile-long journey (the results were published in the September 2009 issue of *Science*). Twenty-nine more eels were released from western Ireland in the autumn of 2008, with tags timed to release in the spring of 2010.

sense is to find the general spawning area. Once they get near it, perhaps temperature, salinity, and the smell of each other come into play, and they begin to aggregate. Of course," he added, "this is all speculation. But it's hard to imagine anything else, really. So *that's* a mystery. How the *heck* do they do that?"

"Is that what you think about when you wake up in the middle of the night?" I asked Mike. "Is that the burning question for you? How they get there?"

"The mysteries I think about every day," Mike said, "are concerned with finding the spawning areas of other species of *Anguilla*—in the South Pacific Ocean, the Indian Ocean, and the Indonesian seas. Katsumi and I are going back to the Indian Ocean next year and the Indonesian seas after that. There are at least two big cruises on the schedule and more planned. We've got the ship time lined up, so we're going to go for it. We're still in the hunt."

⌒

Everyone likes a good treasure hunt. The eel quest requires big toys, major outlays for fuel, a crew, and a staff of scientists. It's a floating community of data collection. But out at sea, Mike says, the purpose of data collection blurs into doubts and uncertainties. With the eel, Mike and his colleagues have found that intuition and spirituality have braided with the science. One of his favorite stories involves his cruise with Katsumi in 1991, when they first discovered the spawning area of the Japanese eel. They were coming up on the last day of the cruise and had captured no eel larvae. So far the expedition had been a bust.

Spirits were low. That night they would make their last pull of the net before heading back.

On the afternoon before that last evening, one of the staff aboard ship conducted an official tea ceremony, being trained and certified in that cultural skill. That evening good fortune came to their nets when they caught not one or two but hundreds of transparent leaf-shaped baby eels, indicating that they were in the vicinity of where adults had spawned. Mike and members of the crew refer to that evening as "magical." They willingly attribute their catch to the spiritual focus provided by the ritual of the tea served by their colleague in a kimono.

"Do you think eels are a spiritual fish?" I asked Mike.

"I'm at my scientific table," he said, laughing. "We might get into that another time. It's not as simple as what people might think."

That's clear. It is hard to evaluate personal experience scientifically. Science is a system to collect and interpret tractable information; obviously, questions related to spiritual issues or personal experience cannot be easily traced and quantified. "A lot of scientists ignore personal experience altogether," Mike said, "largely because it can't be measured. That's not necessarily a mistake if you're doing science. But if you're trying to evaluate life on earth, it probably is."

If you collect information about personal experiences—if you record and interpret the many regenerated stories and recantations—the evidence supports a picture of a far more complex universe.

If an indigenous Pohnpeian tells me he's seen eels come out

of the water and dance in the street on the day before the high chief of his clan dies, or that eels can climb up a tree to take the egg of a fairy tern, who am I to say that doesn't happen? Some might say, "Well, that's not possible." I've come to believe in a system of the universe that can include many things—among them nonphysical existence.

⌒

For me, the truth of the indigenous stories is of no importance. The fact that they exist is. And if they exist in myriad forms in different cultures, that speaks to something essential that I suspect is vitally important to heed. Preserving diversity of fishes or any other type of creature around the world is about preserving the sources of our awe and inspiration. If we lose the creatures that form the foundation of our spiritual systems, if we lose those things that inspire us to be spiritual at all, then we will be lost.

We've been given the gift of inquisitiveness, the capacity to reflect on our own emotions, to create, to imagine. But that gift must be sustained. It can be fed by interpretations of nature already imagined, by books and paintings, by skins and bones of dead animals in natural history museums. But if we can, why not preserve the source as well, to allow people to drink from the original wellspring?

The eel is timeless and vital, a metaphor for the resilience of life itself. To paraphrase what Katsumi wrote in his poem about the eel: Why live, why die? Why do we do these kinds of things? Why do we choose this hard life? Because we have absolutely no choice but to be productive, to be optimistic, to take one step forward, to survive. It's all part of the experience of being alive. To many, Ray Turner—the river rat, the old hair bag—would appear to have a hard life, a lonely life. But to learn about him is to admire him and understand his creed: "Work is a privilege of life." It's the eel in all of us—their relentlessness, as DJ, the Maori bush guide, describes it. Despite the trials, the obstacles, the dams, the diversions, we still make our journeys from birth to death, like the restless traveler, the eel.

ACKNOWLEDGMENTS

This book was many years in the making—about eleven, by my count. (I think it took so long because I've been so fascinated by this fish that I didn't want the investigation to ever end.) I'm hard pressed to remember how exactly it started, but I think it was in conversation with my editor, Larry Ashmead, in Tuscany, Christmas Eve 1998 (as described in the introduction).

Around that time I'd started collecting hand-forged iron eel spears (pictured in the etching on page xi) and was bidding on a Swedish spear on eBay one night when I got an e-mail from a man in France telling me that I was paying too much—that he had a hundred of the same style and could sell me any one of them for less. I told him that I was thinking of writing a book about eels. He said that he was "the eel man in Europe" and that if I was writing about eels I should come to the Basque region of France to see him. That man was Thomas Nielsen, a Dane living in southern France, who is one of the biggest exporters of glass eels from Europe to farms in China. He invited me to spend time with him during the eel season—between December and March—and I flew to Biarritz to do just that. He happened to have one of the finest collections of eel spears in the world, most of them Danish, bought in coastal antique stores on summer jaunts to Denmark with his French wife.

The trip was illuminating for me about the international trade in eels. We drove a shipment of more than two thousand pounds of live glass eels in a tank truck from the village of Charon to a farm in Holland (at $225 a pound, that's nearly half a million dollars' worth of fish). I observed the anxieties of the operation when the load of eels headed for China got sick, and I helped pack another shipment of healthy eels bound for Hong Kong. Later in the trip we went out with glass eel fishermen in the mouth of the Sèvre Niortaise River at night. The boats were fishing the river with no lights (poaching an area of the river that was actually closed), and we had a few close calls and a near collision as the boats dragged fine-mesh nets in the shape of socks on each side of the boat.

After the trip to see Thomas, I officially proposed the book to Larry. Originally, I had envisioned this book as one full of recipes—ways eels were prepared around the world—and of course as a chronicle of the attempts of naturalists and scientists to figure out the elusive life history of eels through the ages.

I spent two weeks with an eel fisherman named Hansa Olofsson on the Ålakust, or Eel Coast, in a region of southern Sweden called Skåne (also known for its apple and wheat production and for Absolut vodka, made in a factory in the town of Åhus, midway up the Ålakust). The drinking of vodka and aquavit was part of the tradition of fall eel dinners, called *alagille*, where eel is served nine to twelve different ways in candlelit fishing shacks on the Baltic Sea. The dinners go late into the night and there is a lot of drinking—it is said that a successful eel dinner is one where the participants don't remember anything the next day. The first person at a dinner to eat enough eel to line the perimeter of the plate, and then make a cross, with eel vertebrae is crowned the eel king.

I also traveled to the town of Comacchio, Italy, where catching and eating eel has been an industry for hundreds of years. The lagoon of Comacchio has been modified at the place where it joins salt water to let young eels in from the Adriatic Sea during spring and let the migrating eels out in the fall. It is very near Venice, in the delta of the Po River. Comacchio is renowned throughout Europe for its eel. A 1954 film, *La Donna del Fiume*, stars Sophia Loren as a woman working in an eel-canning factory.*

But as I did the research, all this wonderful information about the life history of the eel, the culinary aspects of the story (which are cataloged well in a handful of other books about eels), the myriad false guesses about how eels reproduced, and the life and times of Johannes Schmidt began to cave under my interest in the fish as a symbol of mystery for cultures in the Pacific—what the eel means to humans, and to myself. When I started the book I didn't know that eels played any kind of role in the spiritual systems of Pacific peoples, or even that they existed on the islands. Away from the food and natural history and science, the eels led me to an inquiry in my writing and

---

* In Sophia Loren's memoir *Living and Loving*, she tells the story of the first time she ate eel, while working on the movie, which was shot at the Po. "We were served seven different eel dishes that night, each one more delicious than the one before, and if I had not been told I would never have guessed I was eating eel."

painting about myth, about the space between things, and to further investigate our compulsion to control nature through language—how we name and why we name.

⌒

My interests in the beauty and diversity of eel spears led me to Martha's Vineyard, where I met the premier American eel spear collector, Sherman Goldstein. Sherm, of Vineyard Haven, began amassing eel spears in part through his love of the island's history and his passion for striped bass fishing—using live eels for bait, casting them into the surf at night. I enjoyed many a beautiful late fall evening with Sherm, fishing for bass with live eels. I researched the history of and techniques for using eels to catch striped bass, including pickling eel skins and stretching them over wooden lures, a popular fishing method in the chain of islands off the elbow of Cape Cod called the Elizabeths (the most famous for striper fishing being Cuttyhunk Island). I spent time with the other great eel spear collector in the United States, Marcel Salive, who wrote the book on ice-fishing spears. And I learned enough about tools made by coastal blacksmiths on Cape Cod and Long Island, as well as eel spear makers in Denmark and France, to write a small book, or at least a long article, about that.

As anyone who has spent time with me lately knows, I could go on talking about eels forever. So I'll stop now and acknowledge some more people who shared their enthusiasm about eels with me. In my home state of Connecticut, Jimmy Orifice, owner of Jimmy O's tackle shop in Bridgeport, instructed me in fishing for striped bass with eels, as did John Posh and other local fishermen, including Joe Haines (a retired game warden about whom I wrote my second book), who was my first introduction to eating eels. Taylor Hoyt and I have enjoyed many nights casting off the beach in Southport and elsewhere with live eels.

It is David Seidler, a fly-fishing buddy in Santa Monica, who first told me about the importance of eels in Maori culture—and for the introduction to that world I am most grateful.

While I was working on this book, several people who contributed their stories and ideas have passed: Bill Akonga and Kelly Davis, prominent Maori elders or *kaumatua*, as well as Andrew Farmer and Raymond Kitt.

I would like to thank Ray Turner for sharing days at the weir with me, and Don O'Hagan for giving me a key to his cabin across the river from Ray's at Peas Eddy, where I usually stayed on visits there. As with many people I've written about over the years, Ray has become a close friend, and I thank him for his friendship. Stephen Sloan, who has passed, was a great angler and conservationist; he shared with me many an eel conversation, rapt with the wonder and enthusiasm of a child by the mystery of all fishes. Jennifer Reek and Oliver Payne at *National Geographic* helped push my proposal for an article on eels through after two years of pitching (and in editing the story, Ollie helped create the structure this book is built around). The National Geographic Society itself helped mightily with travel, especially to New Zealand and Japan. I thank David Doubilet and Jennifer Hayes for their pleasant company and stories on eel trips, Annie Proulx for erotic eel lore and other obscure tales, John Cook for nights fishing for striped bass at Little Compton and for making the three-legged desk of tupelo and black walnut on which this book was primarily written, and Duncan Galloway for helping build the little shed in my woods where the eel desk sits.

My gratitude to Johannes Schöffmann for taking detours on trout-collecting trips to look for eels in Comacchio, Italy, and in Montenegro; to Joe Dochtermann for bringing me to see the crazy eel sellers at the dawn fish market in Hamburg, Germany; to Jonathan Yang for being candid about how the eel trade works; to Kunio Kadowaki for toting us around Japan and bridging a culture gap; to Stella August for opening my eyes to eels in Maori culture and setting up an itinerary with people who changed the way I look at the world (among them Bill Kerrison, Matt Paku, Hadley Paku, Daniel Joe, and Don Jellyman); to Michael Hopper for sharing his personal history of commercial eel spearing on Cape Cod; to Mike Miller for checking all my science details and suggesting ways to explain things to my readers; to Hal Clifford at *Orion* magazine for helping shape and manage the text and for his encouragement; to Tony Kirk for his assistance at the Center for Contemporary Printmaking in Norwalk, Connecticut, where I produced the etchings for this book; and to Tim Collins for sharing his Adirondack home and providing a quiet place to read and watch the snow fall. Others who read the manuscript and offered comments, or just engaged in helpful discussions with me about eels or rivers or the named and nameless, include Peer Doering, Fred Kircheis,

Acknowledgments

Bill Raynor, Pierre Affre, Barry O'Connor, Terry Holbrook, Susannah Carson, Mei Chin, Annping Chin, Jonathan Spence, James Scott, Harold Bloom, David Skelly, and Elaine Bleakney. Thank you, Hugh Van Dusen and Rob Crawford at Harper-Collins, and Elaine Markson, my agent.

I cannot thank enough my friend Hampton Carey, who read and reread this book with a crucial eye, that of an artist and a historian/scientist, and helped me especially when the book was taking a conceptual turn. And finally, Lauren Hauser has endured more eel talk and eel conversation than anyone imaginable, possibly in the history of humankind, and for that she deserves the greatest helping of eel pie.

BOOKS BY
JAMES PROSEK

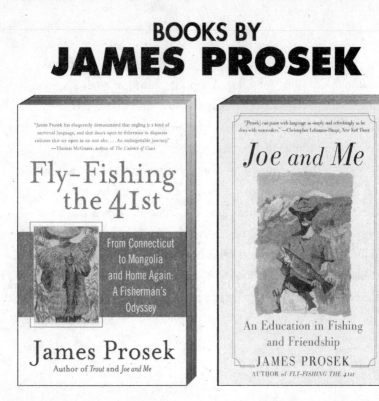

FLY-FISHING THE 41st

From Connecticut to Mongolia and Home Again: A Fisherman's Odyssey

ISBN 978-0-06-055592-4 (paperback)

"The famed fishing writer loops the planet along one of its most interesting latitudinal lines . . . a hit not only with sport fishermen but with anyone who likes to read a well-written adventure."

—*BookPage*

JOE AND ME

An Education in Fishing and Friendship

ISBN 978-0-06-053784-5 (paperback)

"[*Joe and Me* is] about what it means to be tried-and-true; about generosity, responsibility, humor, curiosity, appreciation; about having a warm heart and doing the right thing . . . Prosek describes for us a lost world of sportsfolk—relaxed, comradery, reflective, perceptive—from which he wisely decides to take his cues."

—*Kirkus Reviews*